Mariusz Motyka • Mirosław Skwiot

The Battleship Gneisenau

KAGERO

The contract for construction of the *Panzerschiff E* (*Ersatz Hessen*) was signed with the Deutsche Werke shipyard in Kiel on January 25, 1934. The ship was laid down on February 14 of the same year, but the construction process was halted on July 5, 1934 and the slipway was cleared. In practice, the elements of the hull that had already been built were removed from the slipway. Important, however, was the fact that the contract for the construction of the ship remained valid and it was to be resumed as soon as the modified design had been approved. The most important decision concerning the addition of the third gun turret was made on June 27, 1934 and that was the reason to suspend the construction for some time. However, the 28 cm calibre of the main battery was retained. This decision was taken solely to avoid further delays and the design itself allowed for the subsequent re-armament of the battleship. During the design process, it was recommended that the barbettes were chosen in such a way as to allow for the installation of the larger calibre guns (330, 250 or 380 mm). It was estimated that the re-armament procedure would take approximately 12 to 15 months. The alteration of the number of the main battery guns affected the change of the previous classification from Panzerschiff (armoured ship) into Schlachtschiff (battleship). For the second time the ship was officially laid down on July 6, 1935. The hull was launched on December 8, 1936 and the christening ceremony speech was given by Colonel General Freiherr von Fritsch. The ship's godmother was the widow of the late Julius Maerker – commander of the armoured cruiser *Gneisenau*. The launching ceremony did not go without a minor incident. Due to defective drag weights, the launched hull could not be completely stopped, so it hit the wall of the opposite Hindenburg Embankment. Luckily for the ship, the quay sustained more damage than the hull. Further fitting out work continued according to schedule and on May 21, 1938 the ship was commissioned into the Kriegsmarine. Captain Erich Förster, former commander of the light cruiser *Karlsruhe*, became her commanding officer.

During construction of the battleship a standard transverse and longitudinal layout of the hull braces was used. Individual web frames were set at various distances from 0.75 to 1.5 m, based on the area where they were installed, starting in the stern section, going through the midship towards the bow. The hull was divided into 21 integral watertight compartments, respectively marked with Roman numerals I to XXI. Individual hull compartments had appropriate markings painted on their walls which made it easier for the battleship's crew to orientate inside the hull. There was a autonomous damage control system in each of the compartments which was integrated with the main as well as the auxiliary damage control stations. The armoured citadel located inside the hull protected all the vital installations such as the engine room, magazines, barbettes, etc. New St 42 and St 52 KM steel was used for the hull construction. High-pressure superheated steam turbines were the designed propulsion system for both *Scharnhorst* class battleships. The ship received three sets of turbines delivered by Deschimag A. G., Bremen with the power output of 53 350 SHP each. The steam was generated by 12 high-pressure boilers also manufactured by the Deschimag works. During the sea trials the *Gneisenau* reached the maximum speed of 30.7 knots at the power output of 151 900 HP. Her range at that speed was 2 900 nm, at 19 knots it was 6 200 nm and at 15 knots 8 380 nm. Electrical power was provided by 5 power rooms located in 4 compart-

Table No. 1. General characteristics of the battleship *Gneisenau*.

1	Designed displacement	38.100 t
2	Standard displacement	35.500 t
3	Length (CWL)	226.00 m/ overall 229.80 m *
4	Beam (CWL)	30 m
5	Designed draft	9.9 m
6	Engine room	3 Deschimag A.G. turbines 12 Wagner boilers
7	Power output	160.050 SHP
8	Speed	31.65 knots
9	Armament: Main battery	9 × 28 cm SK C/34
	Secondary battery	12 × 15 cm SK C/28
	Anti-aircraft battery	14 × 10.5 cm SK C/31
		16 × 3.7 cm SK C/30
		10 × 2 cm SK C/30
	Torpedo tubes	6 × 53.3 cm (18 torpedoes)
10	Floatplanes	1 – 3 Ar196A1 – A3
11	Crew	56 officers, 1613 non-commissioned officers and seamen

* after refit 234.90 m

Gneisenau **at the beginning of 1941, seen from the starboard.**

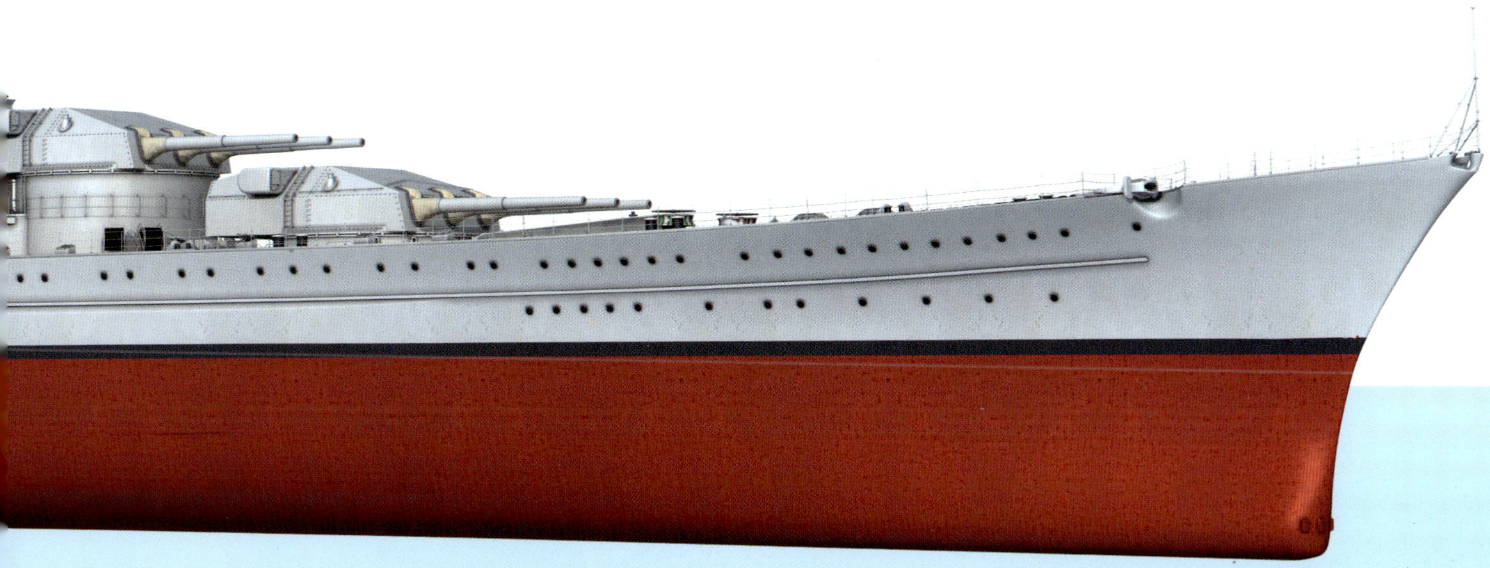

ments. There were four Diesel generators and eight turbogenerators altogether, generating a total of 4 120 kW at 220 V DC.

The *Scharnhorst* class battleships were the first warships of the Kriegsmarine in construction of which new armour plates made of hard Wotan (Wotan Hart) and soft Wotan (Wotan weich) steel were widely used. These materials replaced both nickel steel used so far for the previous *Deutschland* class warships as well as Krupp KNC (Krupp Non-Cemented) steel. Owing to a special welding technology, the new materials allowed for the use of thicker armour in comparison to its weight. They were used for 50 to 150 mm armour plates. The remaining armour was made of face-hardened steel (Krupp Cementiert Panzer – Krupp Cemented Armour), which could be welded. The main side armour stretched on both sides between the frames 32 and 185.7. It extended from 1.7 m below the waterline to 3 m above it. Amidship, it was 170 mm thick below the waterline and 350 mm above it. It the bow area it tapered from 150 to 70 mm and in the stern section respectively from 200 to 170 mm. Above the side armour was the anti-splinter armour made of hard Wotan steel. It was 40-45 mm thick in the citadel area and 35 mm thick in both the bow and stern sections. Internal protection was provided by longitudinal bulkheads. Anti-torpedo bulkhead stretching from frame 32 to 185.7 was made of 45 mm thick soft Wotan steel. Horizontal protection comprised two decks: upper deck and armoured deck. The upper deck that covered the entire hull was made of 50 mm thick hard Wotan steel, while the armoured, deck extending between the forward and aft armoured bulkheads (between the frames 10.5 and 185.7), was made of 80 to 95 mm thick hard Wotan steel. The battleship had three transverse armoured bulkheads, one forward and two aft. The first, 150 mm thick, was installed at the edge of the side armour, near the frame 185.7. The aft bulkhead, similarly to the forward one, was installed at the rear edge of the side armour, near the frame 32. It was made of 200 mm thick Krupp steel plates. The third one, 150 mm thick, was installed right after the steering gear. The armour of the main battery gun turrets varied in thickness from 90 to 360 mm. The face was protected by 360 mm thick armour and a sloped 150 mm thick plate. Side armour plates were 150 to 220 mm[1] thick and the rear armour was 350 mm thick used also to properly balance the turret. The roof was made of 180 mm thick armour plates. The entire turret armour was made of Krupp steel.

Table No. 2. General characteristics of the 28 cm SK C/34 and 15 cm SK C/28 guns.

No.	Gun type	28 cm SK C/34	15 cm SK C/28
1	Calibre	283 mm	149.1 mm
2	Gun weight	53.25 tons	9026/9080 kg
3	Length overall	15.415 mm	8200 mm
4	Barrel length	14.505 mm	7816 mm
5	Chamber length	2619 mm	1152 mm
6	Chamber volume	180 dm³	21.7 dm³
7	Rifling length	11 725 mm	6588 mm
8	Grooves	80 (3.25 mm × 6.72 mm)	44 (1.7 × 6.14 mm)
9	Projectile weight	330 kg	45.3 kg
10	Propellant charge	119 kg RP C/38 (15/4.9)	14.15 kg RPC/38 (7.5/3)
11	Muzzle velocity	890 m/s	875 m/s
12	Working pressure	3200 kg/cm²	3000 kg/cm²
13	Barrel life (effective)	300 rounds	1100 rounds
14	Maximum range	40.930 m at 40°	22.000 m at 35° 23.000 m at 40°

The main battery of the *Gneisenau* was composed of nine quick-firing 28 cm SK C/34 guns manufactured by the Krupp armament works. The C/34 gun for the battleships *Scharnhorst* and *Gneisenau* had the actual calibre of 283 mm and was mounted in Drh L C/28 triple gun turret. In comparison to the previous 28 cm C/34 model, the new gun fired a more elongated projectile, but the propellant charge remained the same as before. The additional weight of the projectile was connected with the use of a new type of warhead, which was assumed to be more effective in piercing armour plates. Early in her career the *Gneisenau* had a catapult installed on turret "C", which considerably reduced its training speed. Consequently, it was removed in February 1940. Then, in the middle of 1942, the *Gneisenau* was disarmed and her "B" and "C" turrets were prepared to be sent to Norway in order to reinforce its coastal defence system. The secondary battery consisted of twelve 150 mm SK C/28 guns installed in four single open MPLC/28 mounts and four twin LC/34 turrets. General characteristics of the 28 cm and 15 cm guns are presented in the Table No. 2

Table No. 3. General characteristics of the 10.5 cm SK C/33 and 37 mm SK C.30 guns.

No.	Gun type	10.5 cm SK C/33	37 mm SK C/30
1	Calibre	150 mm	37 mm
2	Gun weight	4560 kg	243 kg
3	Length overall	6840 mm	3074 mm
4	Barrel length	6348 mm	2960 mm
5	Rifling length	5531 mm	2554 mm
6	Grooves	36 (1.3 × 5.5 mm)	16 (0.55 × 4.76 mm)
7	Projectile weight	15.1 kg	0.748 kg
8	Propellant charge	6.05 kg RPC/40 N (5.5/2.1)	0.365 kg RPC/38 N
9	Muzzle velocity	900 m/s	1000 m/s
10	Working pressure	2850 kg/cm^2	2950 kg/cm^2
11	Effective barrel life	2950 rounds	7500 rounds
12	Maximum range	17.700 m	8500 m/35,7°
13	Maximum ceiling	12.500 m/80°	6800 m/85°
14	Rate of fire	15 rounds/min	30 rounds/min

The heavy anti-aircraft battery of the battleship included 14 quick-firing 10.5 cm SK C/33 guns installed in twin LC 31 gE mounts. Six open mounts were located on the lower level of the forward superstructure near the funnel and a single mount was installed on the aft superstructure in front of turret "C". Medium 37 mm anti-aircraft battery consisted of 16 cannons installed in eight twin C/30 mounts. They were located as follows: four mounts on the lower superstructure deck directly behind turret "B". Next four mounts were installed at the superstructure level on the sides of the aft fire control post. General characteristic of both anti-aircraft battery guns are presented in Table No. 3

The light anti-aircraft battery consisted of 10 single 20 mm MG C/30 cannons. They were replaced by MG C/38 type at the beginning of 1940. During the battleship's entire career the number of the anti-aircraft cannons was frequently changed.

Table No. 4. General characteristics of the 2 cm MG C/30 cannon.

1	Length overall	2300 mm
2	Barrel length	1300 mm
3	Chamber length	140.6 mm
4	Rifling length	1159.4 mm
5	Grooves	8
6	Rifling length in calibres	36 calibres
7	Maximum elevation	+85 degrees
8	Minimum depression	–10 degrees

In the summer of 1941, while the battleship was stationed at Brest, she was equipped with two triple 533 mm torpedo tube mounting, located on the main deck, on both the sides of the aircraft hangar. They were previously used on board the light cruiser *Leipzig*. The ship carried a total of eighteen torpedoes.

The optical fire control system consisted of two 10.5-metre rangefinders installed atop the forward superstructure and on the aft command post. Each of them was mounted on a trainable, gyro-stabilized, armoured post. Both rangefinders directed the main as well as the secondary battery fire. Additionally, each gun turret had its own 10.5-metre rangefinder which allowed for a complete autonomy in case the contact with the fire control post was severed. All the data was transmitted to the fire control centre located under the armoured deck in compartment XIV. It was connected with the main command post through a special communication shaft. In service, after conclusion of Operation "Berlin", the rangefinder was removed from turret "A" and its remaining openings were closed, since it had been constantly flooded by splashes of water. The

The *Gneisenau* – bow view.

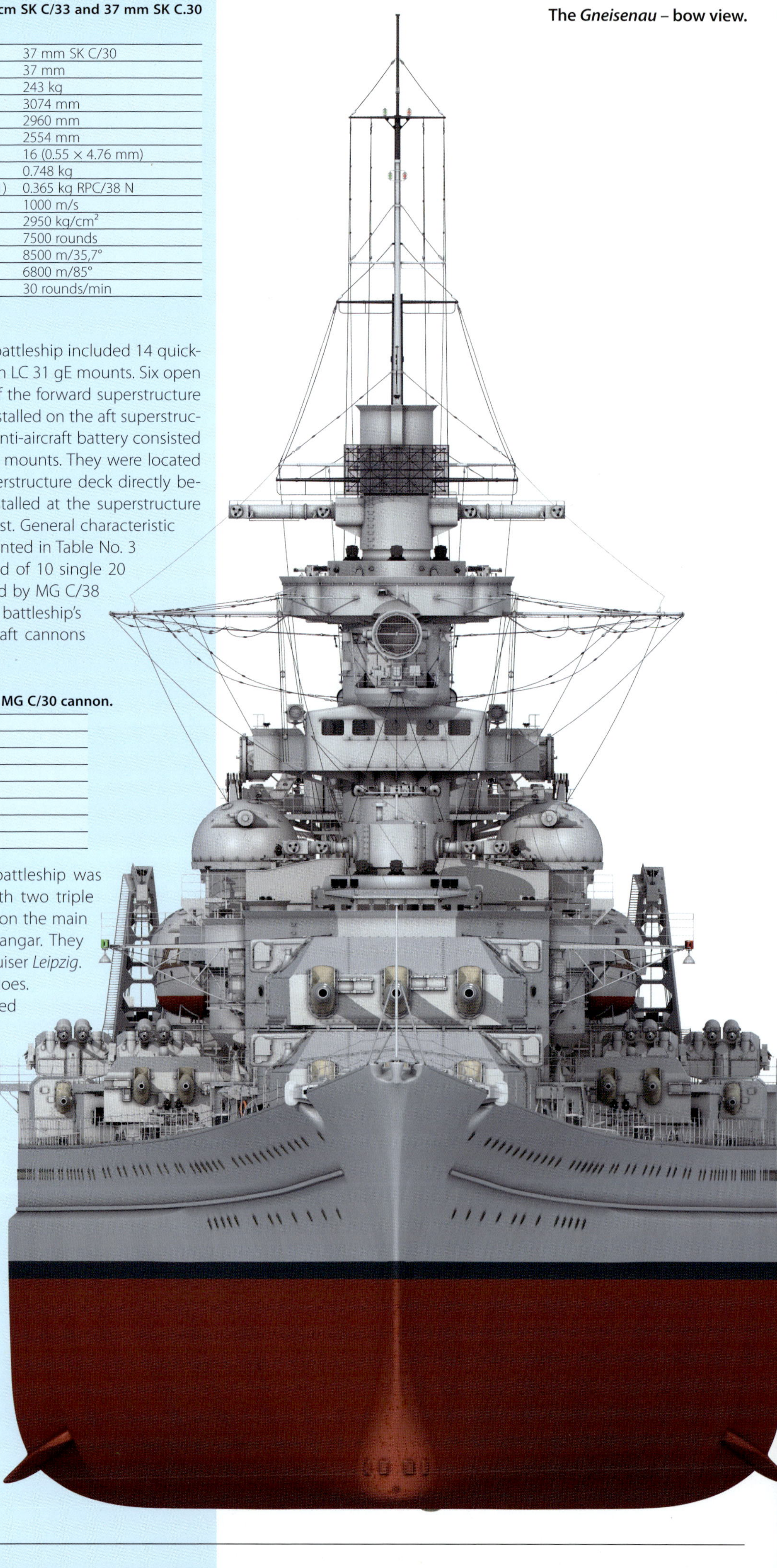

The *Gneisenau* – stern view.

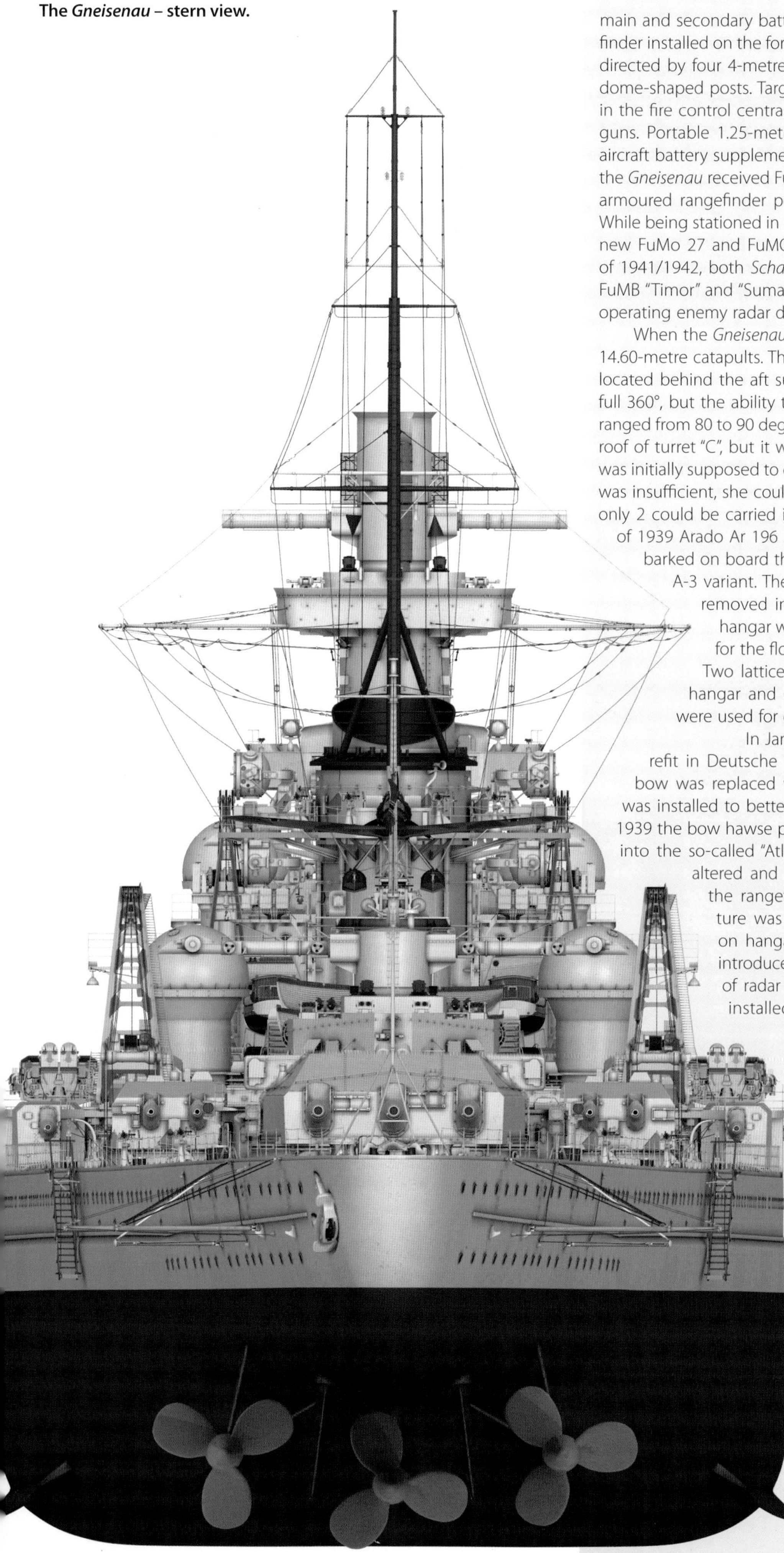

main and secondary battery were also supported by a 6-metre rangefinder installed on the forward fire control post. The anti-aircraft fire was directed by four 4-metre SL-6 rangefinders installed in gyro-stabilized dome-shaped posts. Target data was transmitted to one of the rooms in the fire control central. From there it was sent directly to individual guns. Portable 1.25-metre base rangefinders used by the light anti-aircraft battery supplemented the optical equipment. In October 1939 the *Gneisenau* received FuMO 22 radar. Its antenna was installed on the armoured rangefinder post located atop the forward superstructure. While being stationed in Brest, in the summer of 1941 the ship received new FuMo 27 and FuMO 21 radar equipment. Moreover, at the turn of 1941/1942, both *Scharnhorst* class battleships were equipped with FuMB "Timor" and "Sumatra" antennas that were capable of detecting operating enemy radar devices.

When the *Gneisenau* entered service, she was equipped with two 14.60-metre catapults. The first one was installed on an 8-metre tower located behind the aft superstructure. Admittedly, it could be trained full 360°, but the ability to launch a floatplane was reduced to ¼ and ranged from 80 to 90 degrees. The second catapult was installed on the roof of turret "C", but it was removed in February 1940. The battleship was initially supposed to carry 4 floatplanes, but since the hangar space was insufficient, she could theoretically embark 3 planes, but in reality only 2 could be carried including one on the catapult. In the middle of 1939 Arado Ar 196 A-1 of the 1st Bordfliegerstaffel 196 were embarked on board the battleship. They were later replaced by the A-3 variant. The small hangar located near the catapult was removed in 1939. Only in the autumn of 1940, a new hangar was constructed in that section, which allowed for the floatplanes to be launched directly from within. Two lattice hoisting cranes located on the sides of the hangar and one boom installed at the side of turret "C" were used for operating floatplanes.

In January 1939 the battleship underwent her first refit in Deutsche Werke shipyard in Kiel. The previous clipper bow was replaced with the new stem. Moreover, a funnel cap was installed to better facilitate the flow of exhaust fumes. In May 1939 the bow hawse pipes were modified and the stem was rebuilt into the so-called "Atlantic bow". The funnel cap was also slightly altered and the radio topmast (Funkstenge) installed on the rangefinder located atop the forward superstructure was replaced by a larger one. The smaller built on hangar was removed. Further modifications were introduced in October 1939 including the installation of radar equipment with FuMO 22 mattress antenna installed atop the forward 10.5-metre rangefinder. While the ship was dry-docked, in December 1939, the admiral's bridge of the forward superstructure was enclosed and she also received the degaussing cable installation (MES). In February 1940 the catapult installed atop turret "C" was removed at the Kriegsmarine Werf shipyard in Wilhelmshaven. At the end of 1940, a single Flakvierling C/35 cannon mount was installed on board the battleship for artillery test purposes and since January 14, 1941, all 20 mm MG C/30 cannons were being replaced by the new MG C/38 mounts. Also, a Flakvierling C/35 mount was installed on a platform behind the funnel. In the summer of 1941, while the battleship was stationed at Brest, FuMO 27 radar was installed atop the aft rangefinder. During the installation of a new hangar amidship, the Flakvierling platform was removed and the mount itself was installed on a platform atop the hangar. Before the beginning of Operation "Cerberus" the existing anti-aircraft armament was reinforced. The admiral's bridge was also rebuilt by shortening its wings by half.

Operational history

The outbreak of World War II found the *Gneisenau* in Brunsbüttel, where on September 4, 1939, she was unsuccessfully attacked by British bombers. Four days later, along with the *Scharnhorst*, she sailed through the Nord-Ostsee channel to the Baltic Sea, where she began artillery drills. On completion of those, on October 5, 1939, she became the flagship of the Commander-in-Chief of the Kriegsmarine, Admiral Hermann Boehm. Her first combat mission was a sortie with the light cruiser *Köln* and destroyers against the ship routes between Great Britain and Norway. It was soon over and the ships arrived in Kiel. On the next mission the *Gneisenau* departed with the light cruisers *Köln* and *Leipzig*, as well as her sistership the *Scharnhorst*. During this sortie the force under command of Vice Admiral Wilhelm Marshall sank the British armed merchant cruiser HMS *Rawalpindi*. The battleship was damaged in the storm. The hull was leaking and the ship was taking on considerable amounts of water. The situation was dangerous enough to abort the mission. The necessary repairs were made in Kiel and then the ship was redeployed to the Baltic Sea to undergo the sea trials. In the middle of January 1940 the *Gneisenau* returned to Kiel, where she was being prepared for another commerce raiding sortie in the Atlantic Ocean. This action, known as Op-

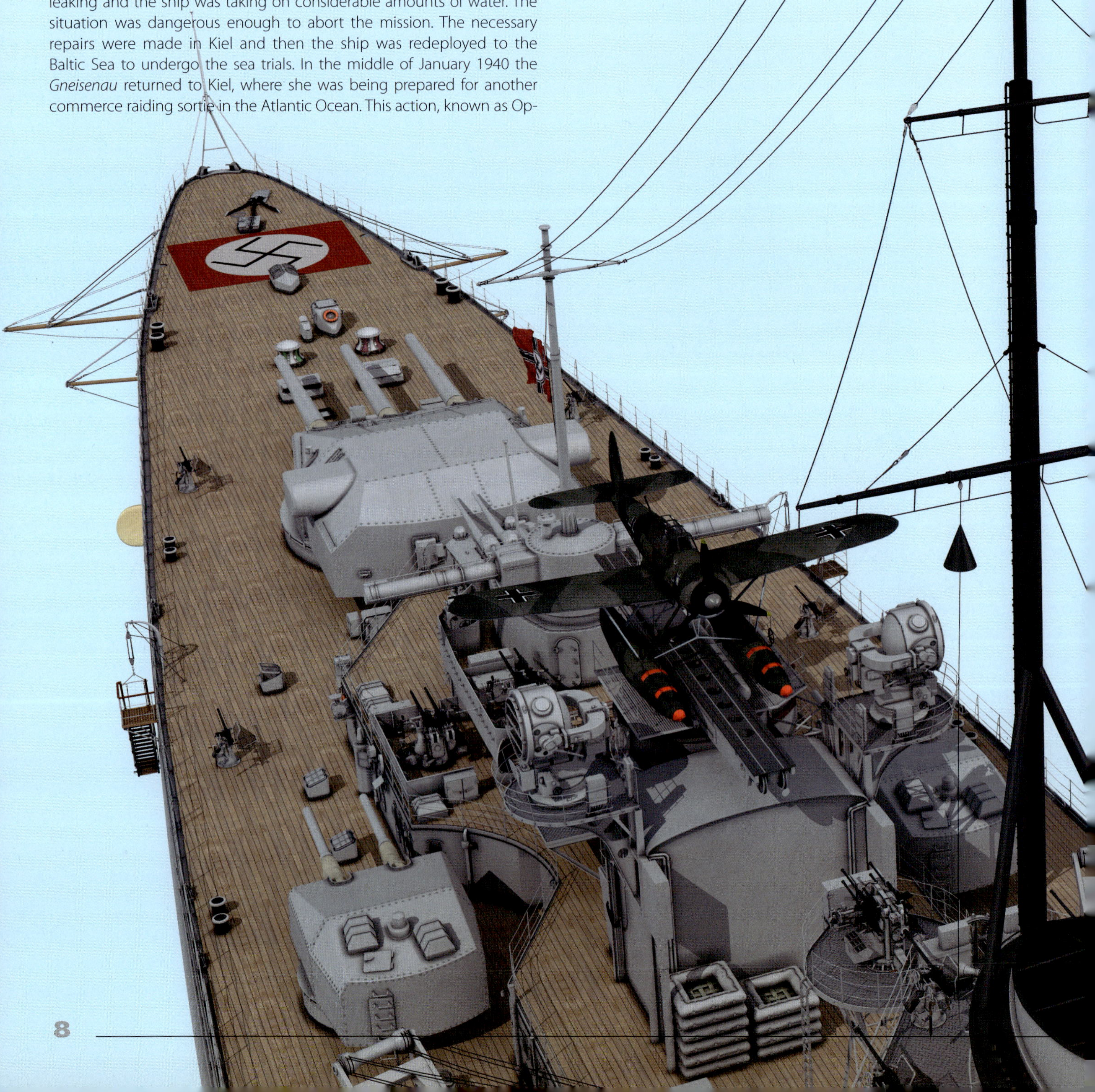

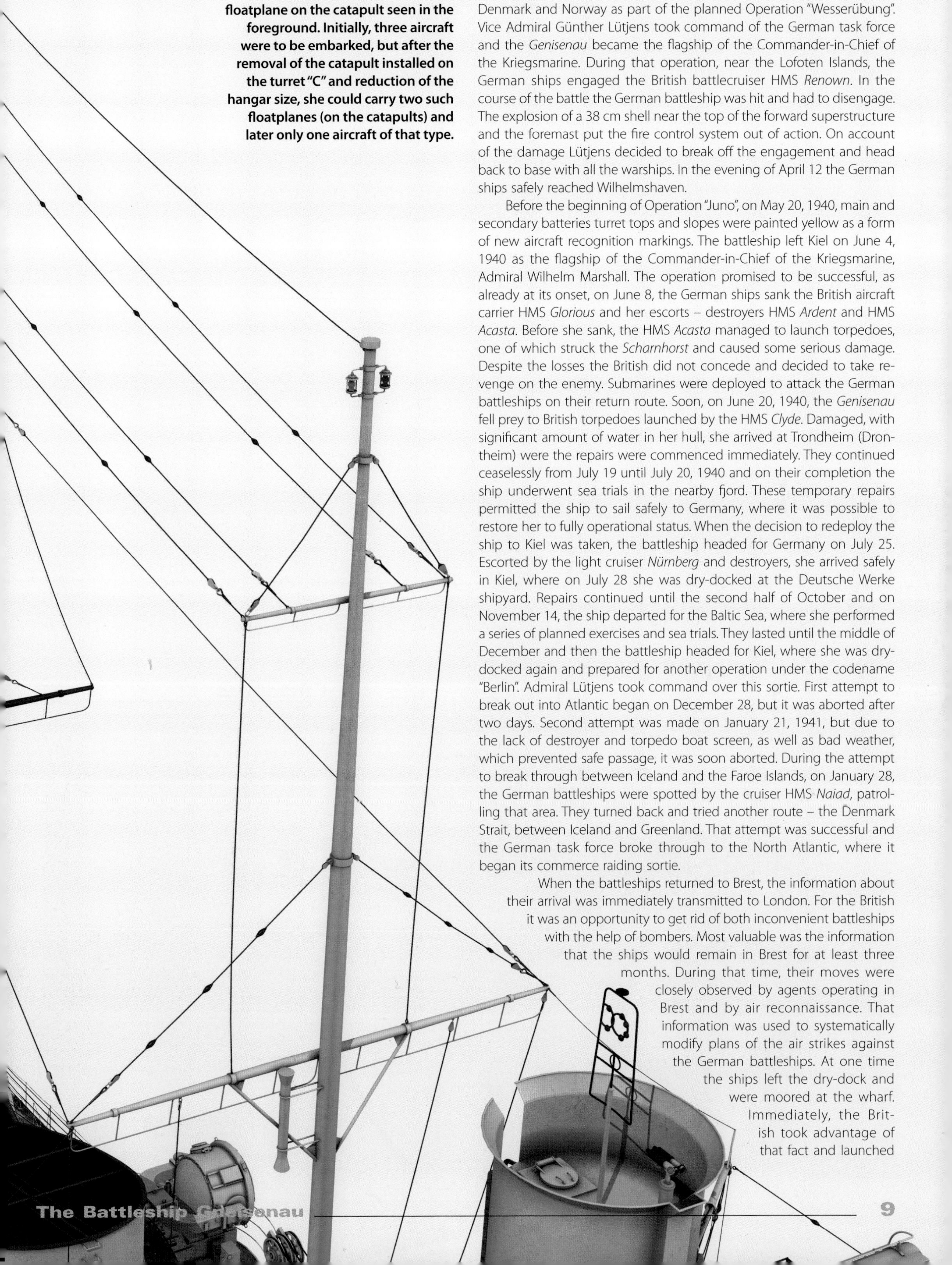

Stern view of the battleship *Gneisenau* with her Arado Ar 196 floatplane on the catapult seen in the foreground. Initially, three aircraft were to be embarked, but after the removal of the catapult installed on the turret "C" and reduction of the hangar size, she could carry two such floatplanes (on the catapults) and later only one aircraft of that type.

eration "Nordmark" was a failure. Later, along with the other warships of the Kriegsmarine, the battleship was being prepared for the invasion of Denmark and Norway as part of the planned Operation "Wesserübung". Vice Admiral Günther Lütjens took command of the German task force and the *Genisenau* became the flagship of the Commander-in-Chief of the Kriegsmarine. During that operation, near the Lofoten Islands, the German ships engaged the British battlecruiser HMS *Renown*. In the course of the battle the German battleship was hit and had to disengage. The explosion of a 38 cm shell near the top of the forward superstructure and the foremast put the fire control system out of action. On account of the damage Lütjens decided to break off the engagement and head back to base with all the warships. In the evening of April 12 the German ships safely reached Wilhelmshaven.

Before the beginning of Operation "Juno", on May 20, 1940, main and secondary batteries turret tops and slopes were painted yellow as a form of new aircraft recognition markings. The battleship left Kiel on June 4, 1940 as the flagship of the Commander-in-Chief of the Kriegsmarine, Admiral Wilhelm Marshall. The operation promised to be successful, as already at its onset, on June 8, the German ships sank the British aircraft carrier HMS *Glorious* and her escorts – destroyers HMS *Ardent* and HMS *Acasta*. Before she sank, the HMS *Acasta* managed to launch torpedoes, one of which struck the *Scharnhorst* and caused some serious damage. Despite the losses the British did not concede and decided to take revenge on the enemy. Submarines were deployed to attack the German battleships on their return route. Soon, on June 20, 1940, the *Genisenau* fell prey to British torpedoes launched by the HMS *Clyde*. Damaged, with significant amount of water in her hull, she arrived at Trondheim (Drontheim) were the repairs were commenced immediately. They continued ceaselessly from July 19 until July 20, 1940 and on their completion the ship underwent sea trials in the nearby fjord. These temporary repairs permitted the ship to sail safely to Germany, where it was possible to restore her to fully operational status. When the decision to redeploy the ship to Kiel was taken, the battleship headed for Germany on July 25. Escorted by the light cruiser *Nürnberg* and destroyers, she arrived safely in Kiel, where on July 28 she was dry-docked at the Deutsche Werke shipyard. Repairs continued until the second half of October and on November 14, the ship departed for the Baltic Sea, where she performed a series of planned exercises and sea trials. They lasted until the middle of December and then the battleship headed for Kiel, where she was dry-docked again and prepared for another operation under the codename "Berlin". Admiral Lütjens took command over this sortie. First attempt to break out into Atlantic began on December 28, but it was aborted after two days. Second attempt was made on January 21, 1941, but due to the lack of destroyer and torpedo boat screen, as well as bad weather, which prevented safe passage, it was soon aborted. During the attempt to break through between Iceland and the Faroe Islands, on January 28, the German battleships were spotted by the cruiser HMS *Naiad*, patrolling that area. They turned back and tried another route – the Denmark Strait, between Iceland and Greenland. That attempt was successful and the German task force broke through to the North Atlantic, where it began its commerce raiding sortie.

When the battleships returned to Brest, the information about their arrival was immediately transmitted to London. For the British it was an opportunity to get rid of both inconvenient battleships with the help of bombers. Most valuable was the information that the ships would remain in Brest for at least three months. During that time, their moves were closely observed by agents operating in Brest and by air reconnaissance. That information was used to systematically modify plans of the air strikes against the German battleships. At one time the ships left the dry-dock and were moored at the wharf. Immediately, the British took advantage of that fact and launched a torpedo bombers strike against the *Gneisenau*. It was truly a master stroke, since the Germans did not expect a torpedo attack against a vessel moored in the narrow inner basin. Admittedly, such a possibility had been discussed, but there was not enough time to lay the anti-torpedo nets. A daring attack made by Flying Officer Kenneth Campbell scored a torpedo hit on the *Gneisenau*. Four days later the British launched another air raid and the ship was hit by bombs, which put her out of action for a few months. On April 14, 1941, Admiral Lütjens made an inspection of the *Gneisenau*, and it was estimated that all the necessary repairs would last at least until autumn.

The *Gneisenau* remained in Brest until the end of the year, when on November 5, 1941, she was finally made fully operational. Her further fate is connected with Operation "Cerberus". In February 1942, along with her sistership the *Scharnhorst* and the heavy cruiser *Prinz Eugen*, she dashed through the English Channel to Germany, under the nose of the British. During that sortie, on February 12, she hit a mine, but despite the damage she was safely moored at Brunsbüttel on the next day. Within the next days, she sailed through the canal and on February 15 was dry-docked at the Deutsche Werke shipyard in Kiel with ammunition still on board. Restoring her to fully operational status was a priority and the repairs were scheduled to conclude on March 6, 1942. Therefore, permission to keep the ammunition on board was granted on one condition – shell fuses had to be removed. As it soon turned out, it did not protect the ship from danger. In the meantime, the British were not idle and on the night of February 26-27, 1942 launched an air raid against the battleship. The *Gneisenau* was hit by a bomb, which exploded in the bow section. Hot gases and splinters ignited propellant charges in turret "A" which cased an explosion. Consequently, the entire bow section of the battleship was demolished, which put her out of action for at least a few months.

The damage was not fatal, and it was decided to rebuild the ship in order to arm her with the new 38 cm guns according to design specifications approved by the OKM (Oberkommando der Marine – Naval High Command) on July 21, 1942. The approved design alteration also included lengthening of the bow section by 10 metres to the total of 245 metres. According to calculations, standard displacement would be approximately 33 510 tons, while operational displacement would reach 40 720 tons. It was assumed that the 15 cm and 10.5 cm guns would be replaced by a battery of dual purpose 12.8 cm SK C/42 guns. The design also stipulated the modification of the superstructure, including construction of the mainmast identical to that of the *Scharnhorst* and removal of the tall funnel mast. The ship was decommissioned on July 1, 1942 to be thoroughly rebuilt and rearmed. However, after the failure of Operation "Regenbogen" Adolf Hitler decided to decommission all capital ships. Therefore, the conversion work on the *Gneisenau* was stopped and both remaining main battery turrets were removed and installed in Norway as shore batteries. The ship remained in Gotenhafen (now Gdynia) serving as the accommodation hulk and floating warehouse.

The *Gneisenau* spent her final moments in harbour basin III. She was there on March 23, 1945, when the destroyer *Z 31* towed her to the entrance of the harbour, where the Germans planned to scuttle her and thus block it. The operation began on March 27, 1945 by opening the Kingston valves. Additionally, the destroyer *Z 31* launched torpedoes, two of which exploded on the port side in the engine room area. Water rushed through the openings, and although the explosion torn the side armour plates and destroyed a number of compartments, it failed to pierced the engine room bulkhead. Only the detonation of explosives finally flooded the engine room. For a few years after the war, no salvage work was attempted. The contract between the harbour authority and the Polish Ship Salvage (Polskie Ratownictwo Okrętowe) for refloating and removal of the wreck was not signed until in April 1950. The work commenced on April 28, 1950, but only on September 7, 1951 the sealed hull was refloated. Then it was towed away from the breakwater to clear the harbour entrance. At that time moving the wreck towards the wharf turned out to be impossible because the Dredging and Underwater Works Company (Przedsiębiorstwo Robót Czerpalnych i Podwodnych) had just begun dredging the section of the channel leading from the wreck to the wharf. At 10 metres the ship's draft was too big, so the entire towing operation was postponed until the end of April. At the beginning of 1952, a meeting was called at the Wreckage Removal Company (Przedsiębiorstwo Demontażu Wraków) to discuss the removal of the *Gneiseanu*'s armour plates. Relatively soon they were removed and what was most important, by July 1952, all the machinery and equipment with a total weight of around 2500 tons had been removed from the wreck. The final sections of the battleship's floor plates were ceremonially scraped on July 15, 1953, a few days before the scheduled date set to July 22, 1953.

Endnotes

[1] Various German language sources provide conflicting data concerning the armour thickness.

Top view of the battleship's midship section with the exact location of secondary and anti-aircraft batteries clearly visible.

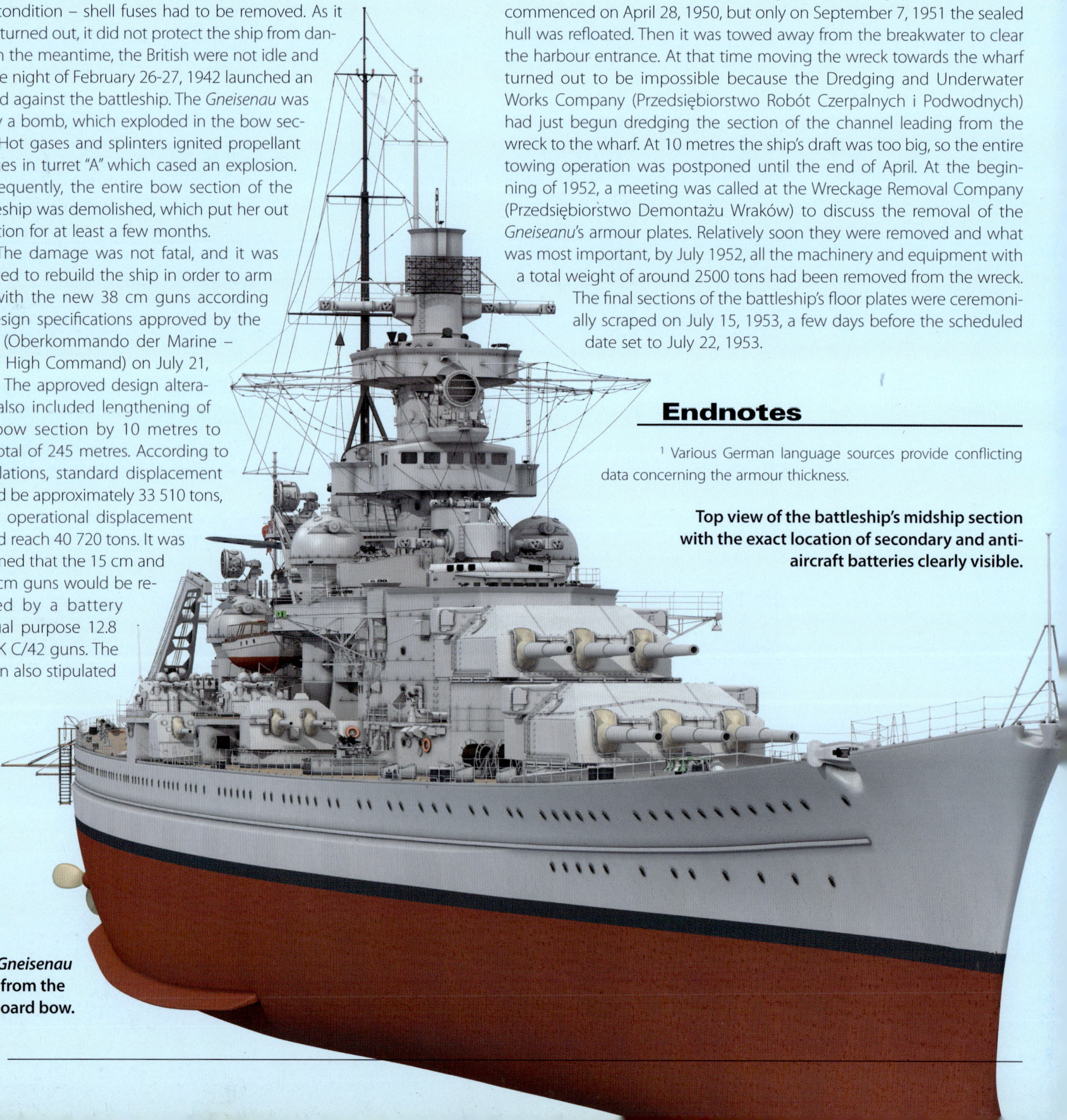

The *Gneisenau* seen from the starboard bow.

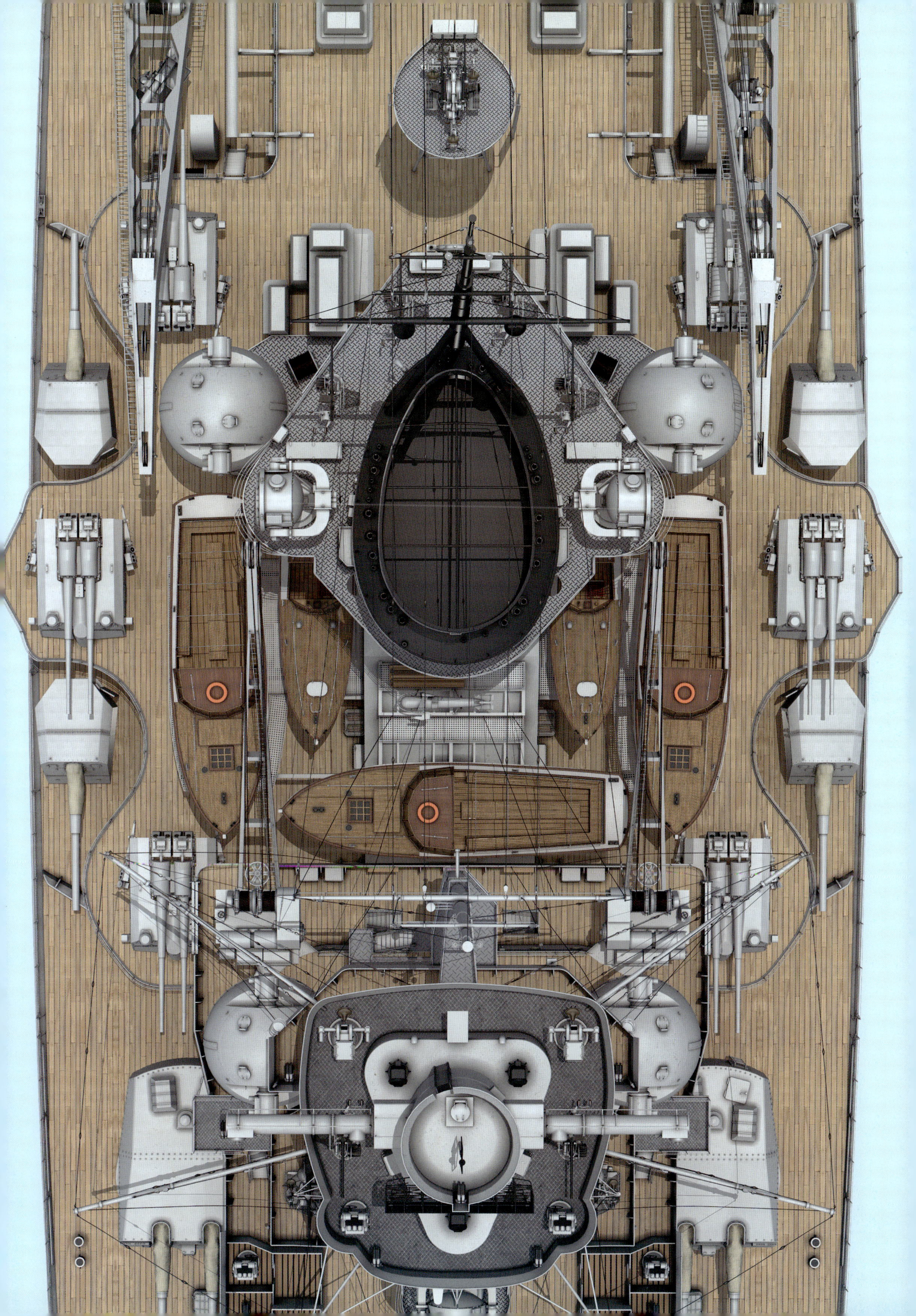

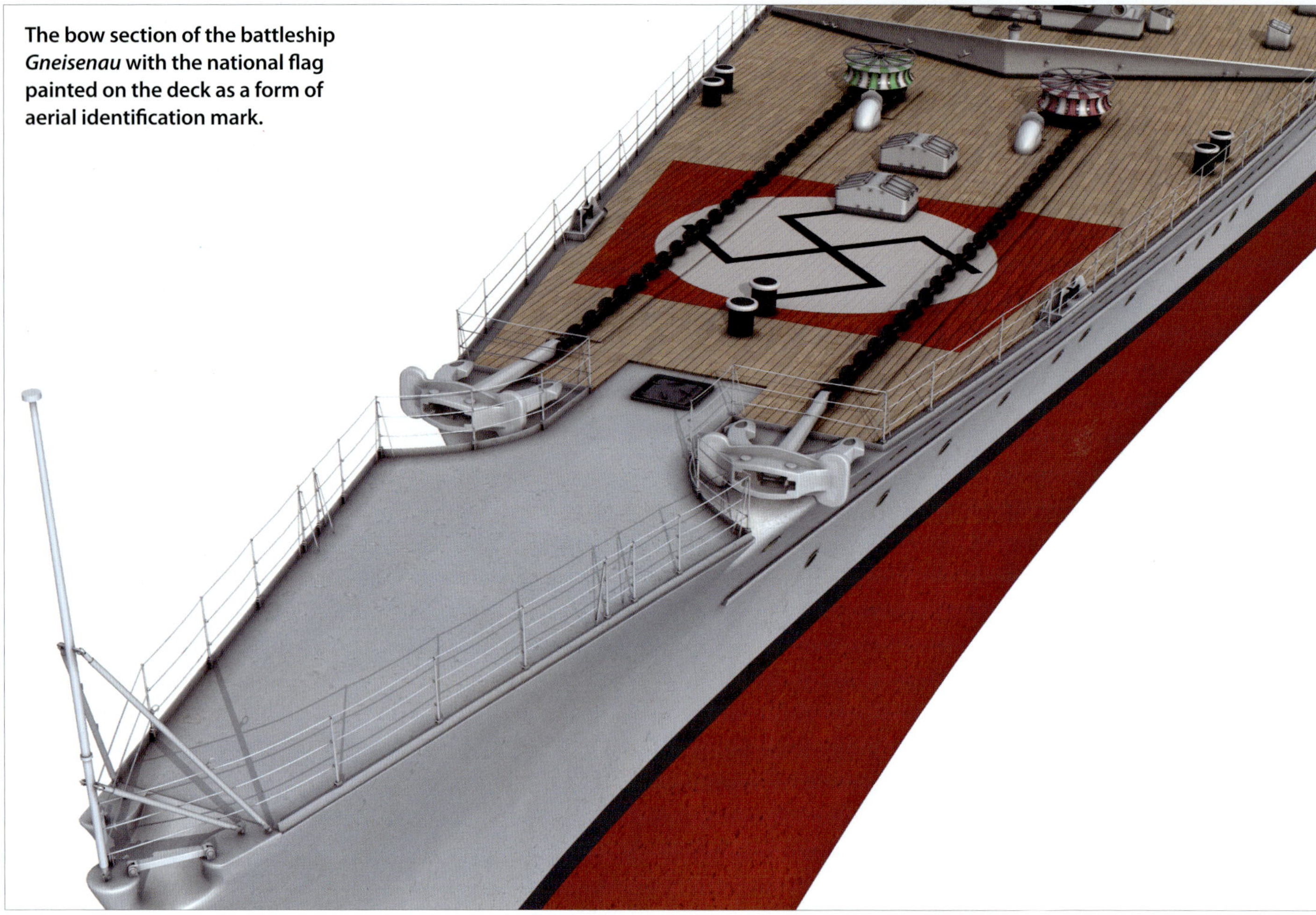

The bow section of the battleship *Gneisenau* with the national flag painted on the deck as a form of aerial identification mark.

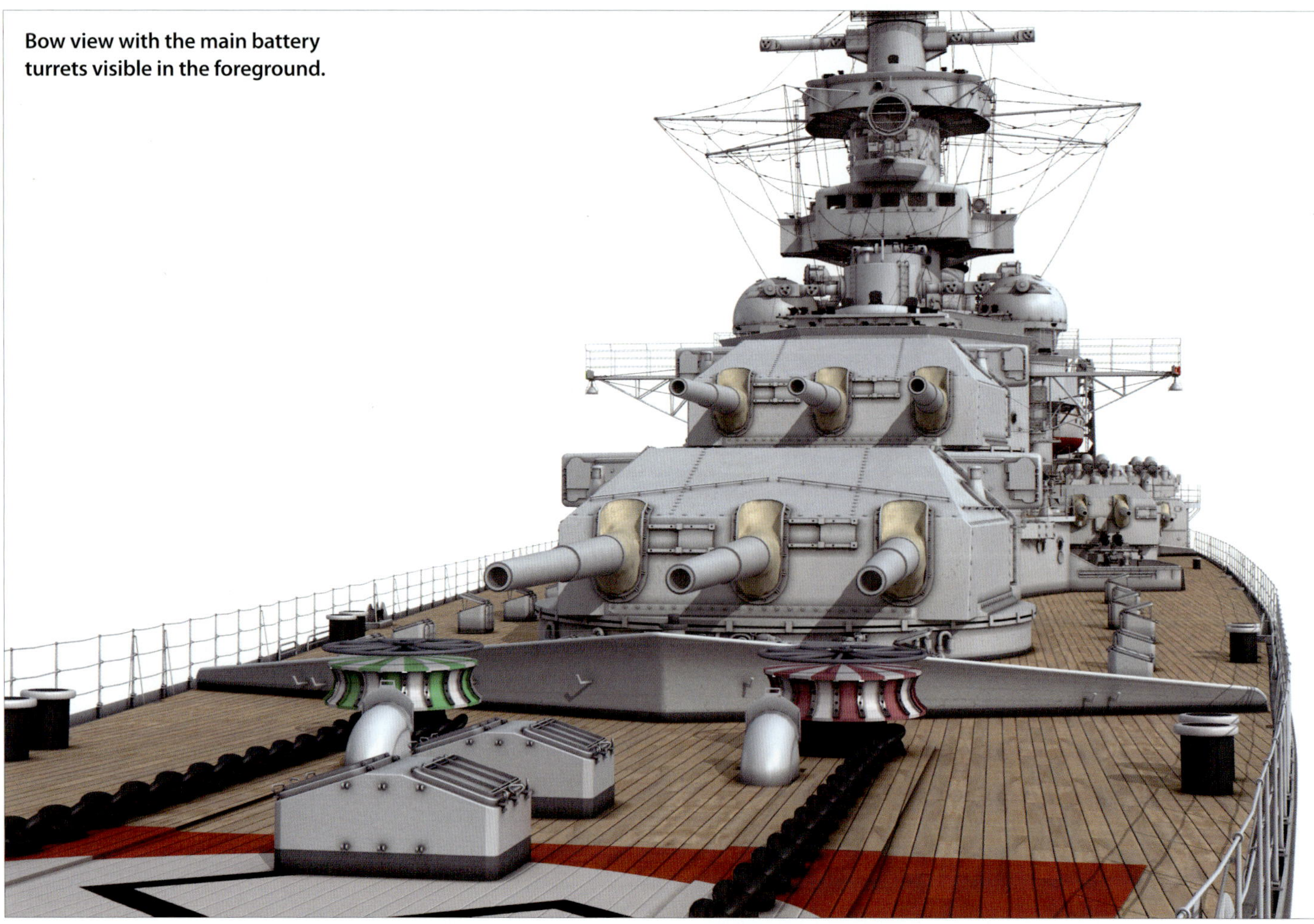

Bow view with the main battery turrets visible in the foreground.

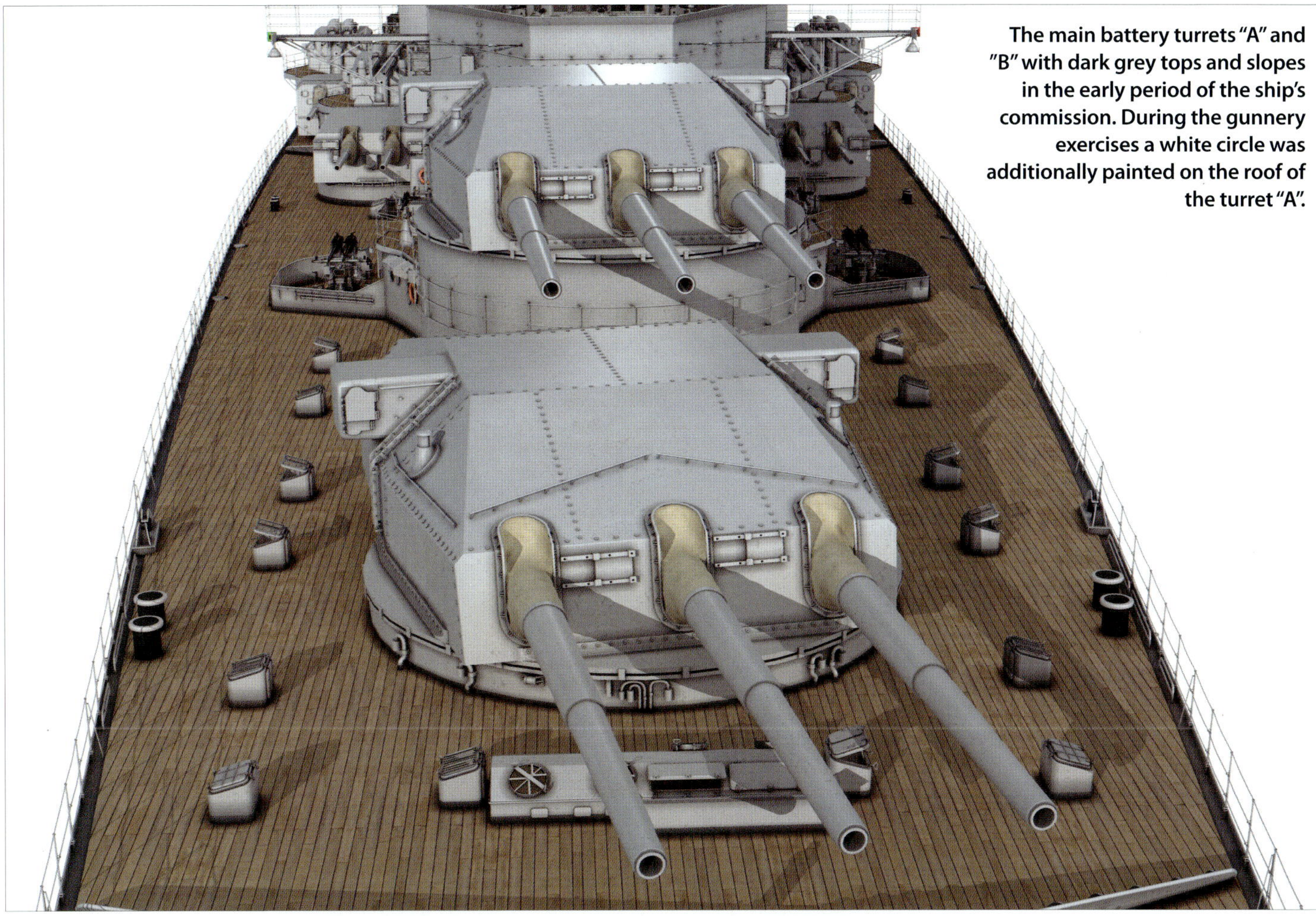

The main battery turrets "A" and "B" with dark grey tops and slopes in the early period of the ship's commission. During the gunnery exercises a white circle was additionally painted on the roof of the turret "A".

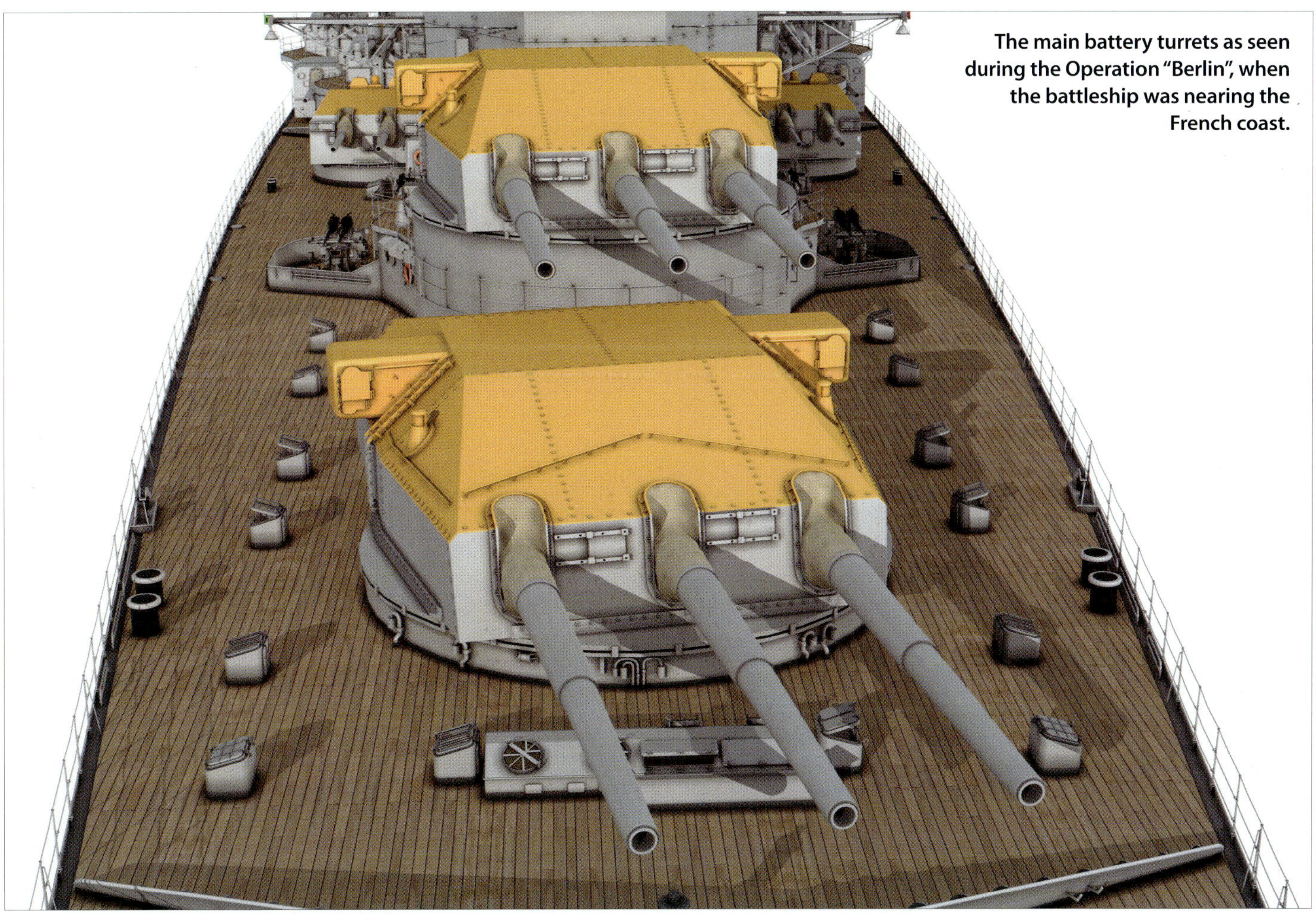

The main battery turrets as seen during the Operation "Berlin", when the battleship was nearing the French coast.

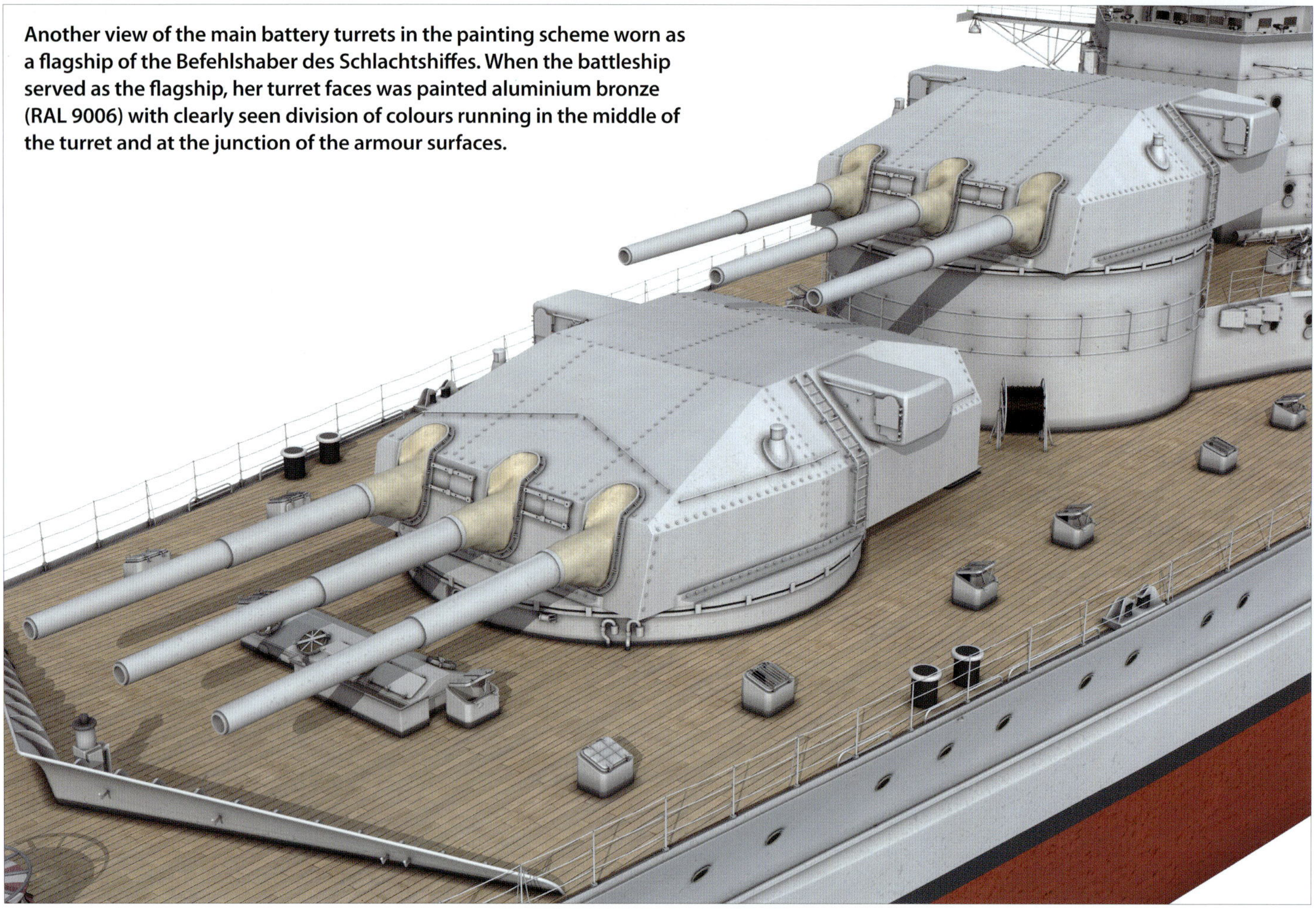

Another view of the main battery turrets in the painting scheme worn as a flagship of the Befehlshaber des Schlachtshiffes. When the battleship served as the flagship, her turret faces was painted aluminium bronze (RAL 9006) with clearly seen division of colours running in the middle of the turret and at the junction of the armour surfaces.

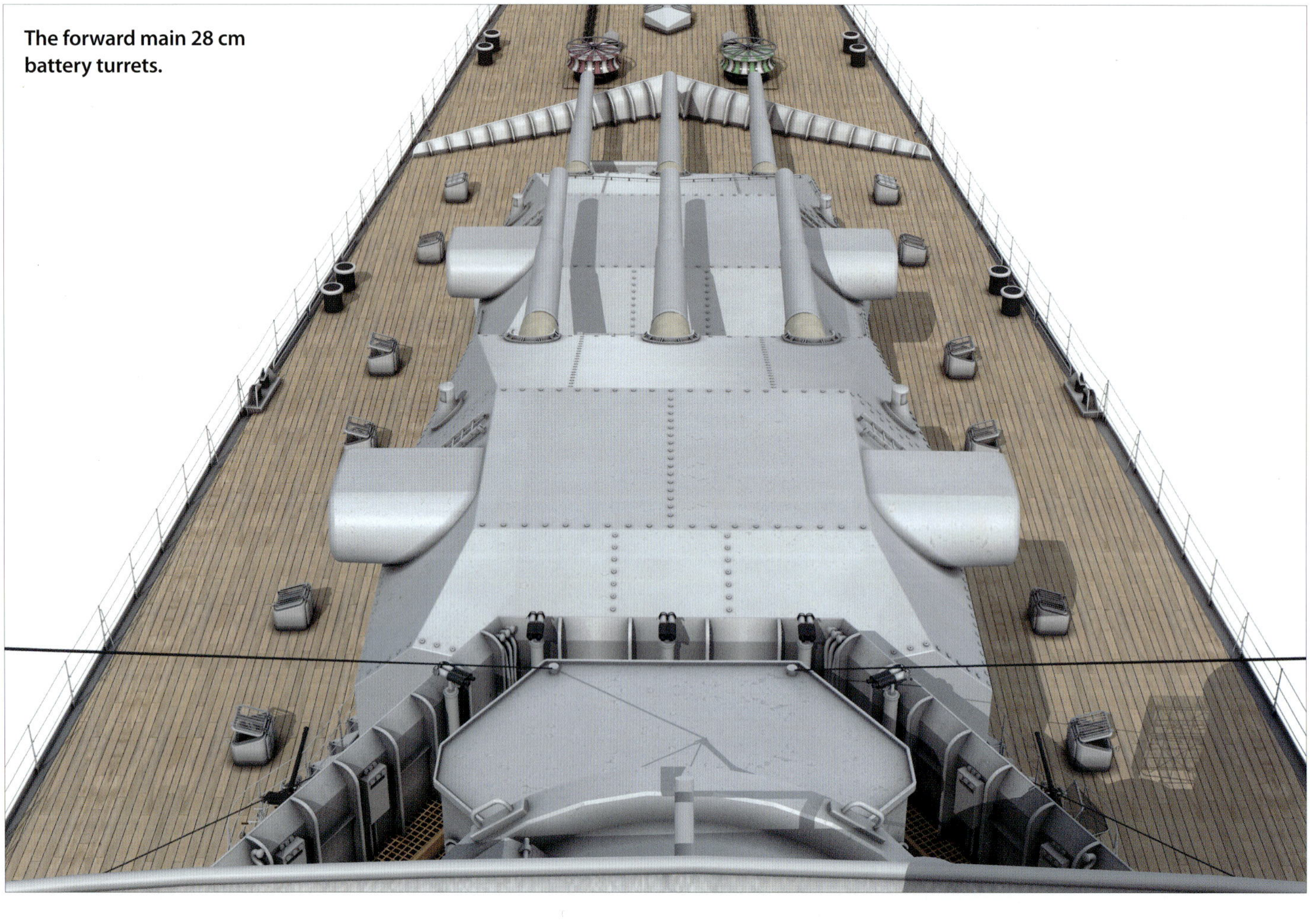

The forward main 28 cm battery turrets.

The main battery turret "B" of the battleship *Gneisenau* with the forward superstructure visible behind it.

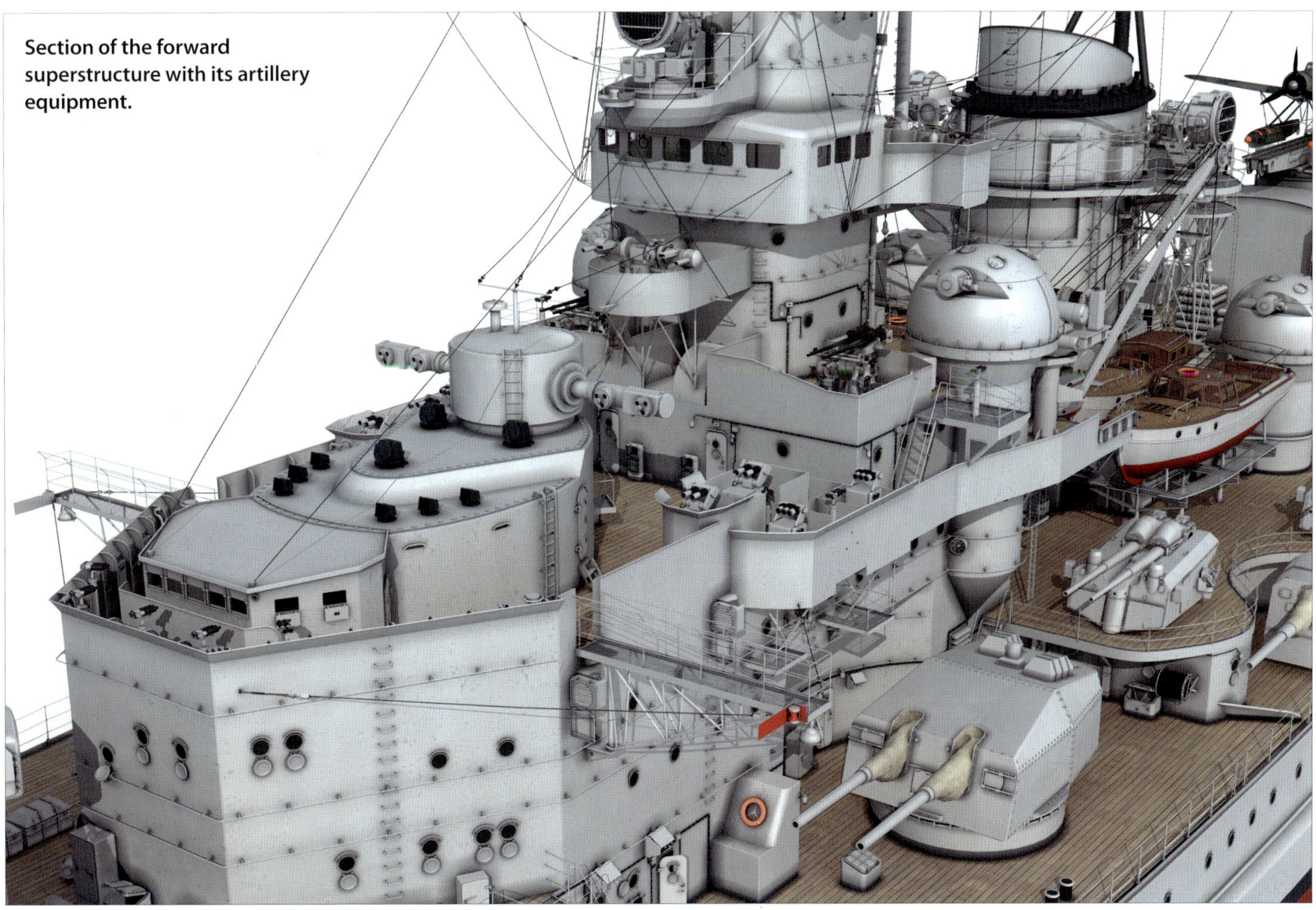

Section of the forward superstructure with its artillery equipment.

The forward superstructure seen from the port.

The forward superstructure seen from the port. Admiral's bridge and the top platform with its braces are clearly visible.

Port side of the forward superstructure.

Forward 15 cm twin secondary battery LC/34 turret.

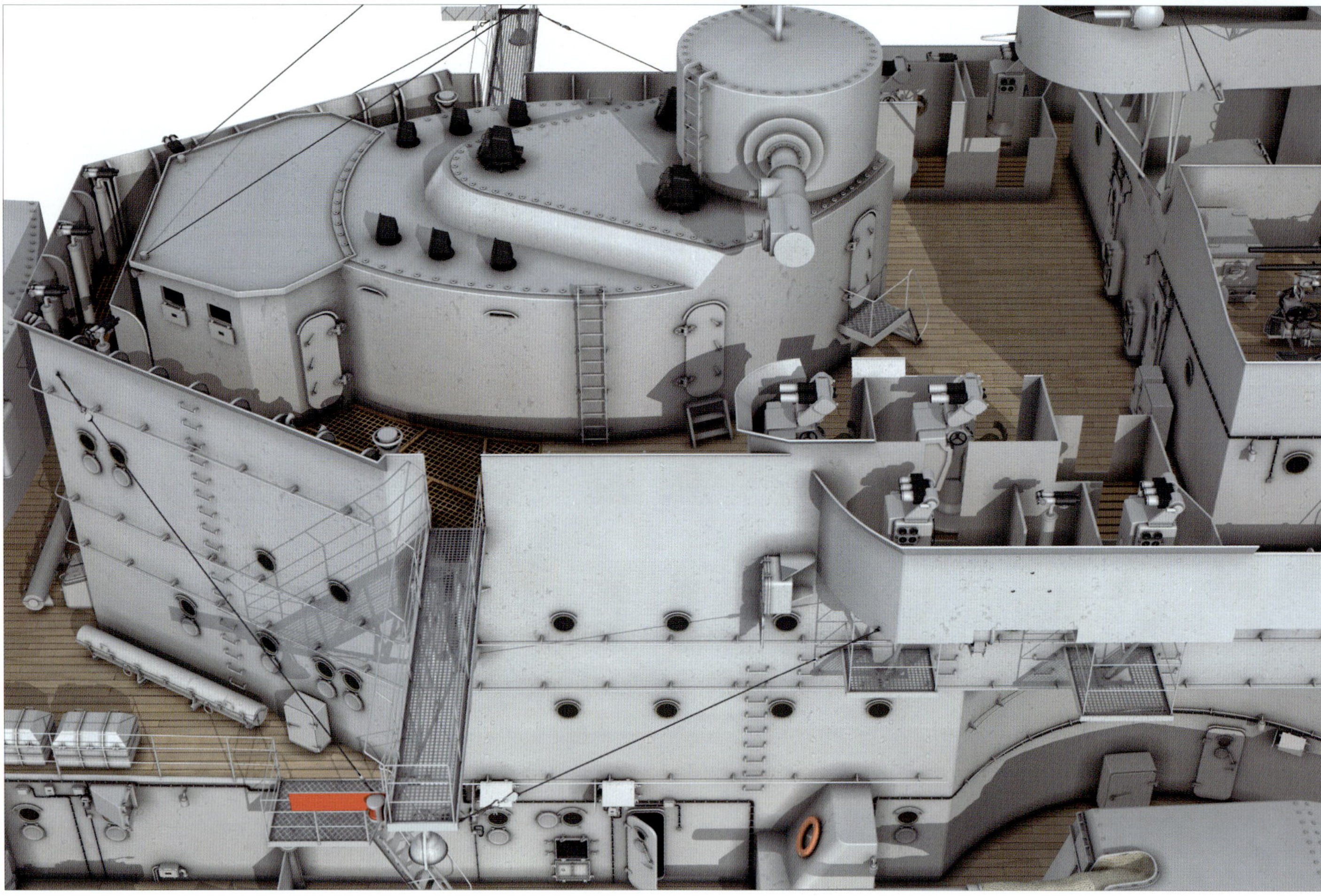
The 6-metre rangefinder installed on top of the forward fire control post.

The bow section of the *Gneisenau*.

The Admiral's bridge.

The FuMO 22 radar antenna installed atop the 10.5-metre rangefinder.

The forward superstructure. The SL-6 anti-aircraft battery rangefinder is visible on the right and one of the ship's searchlights can be seen above on the left. The 10.5-metre rangefinder with the FuMO 22 radar antenna are on the top platform.

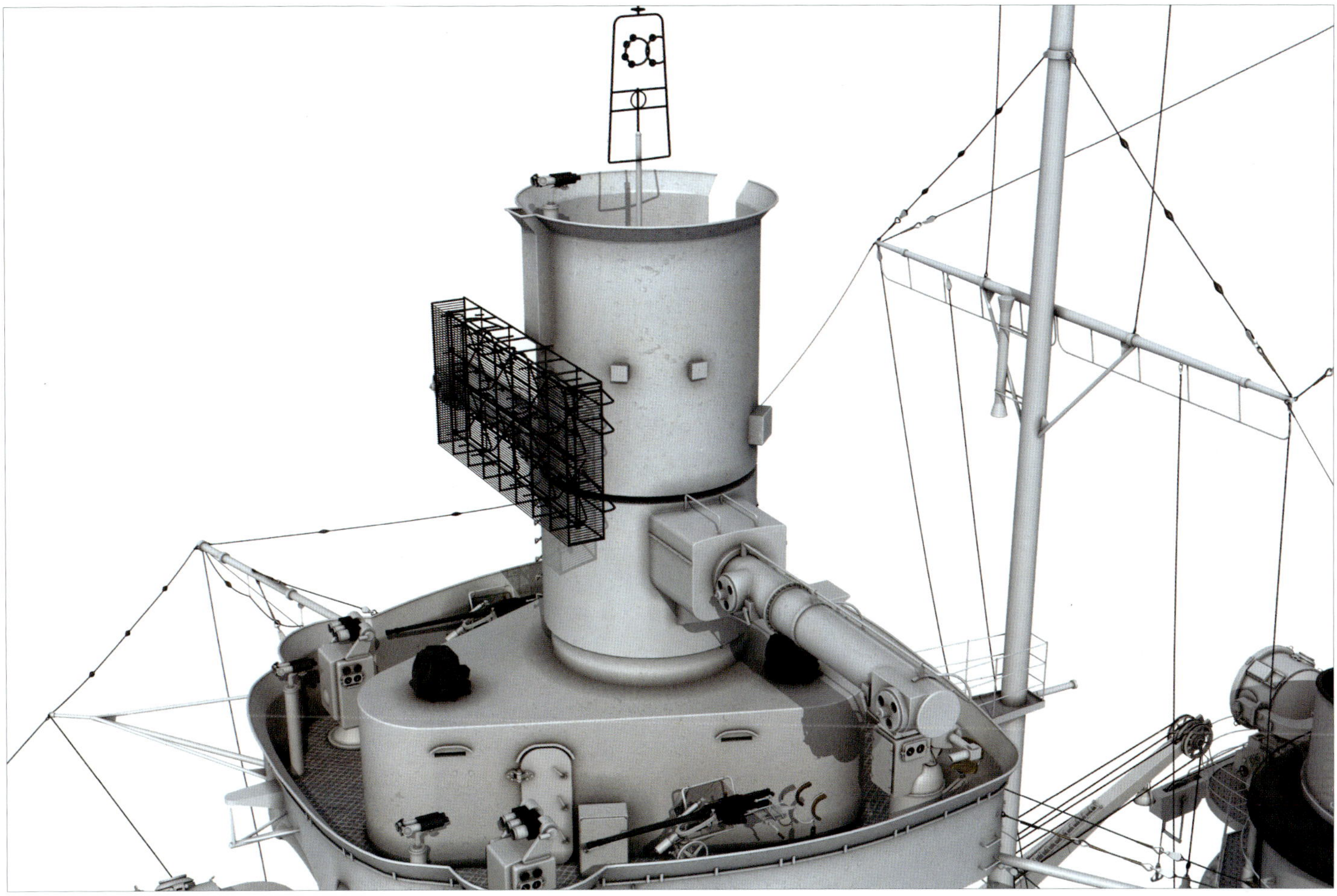

Fire control post on the top platform.

The midship section of the *Gneisenau*.

The forward superstructure with its fittings seen from the port astern.

The top platform braces.

The battleship's forward superstructure.

The funnel with its fittings seen from the port.

The Admiral's bridge and the entrance door to the superstructure.

The midship section of the battleship with its fittings as seen during the Operation "Berlin".

Details on the port side of the funnel.

The bow section of the *Gneisenau* seen from the port.

The battleship's midship with its fittings.

The battleship's fittings located around the funnel.

The SL-6 rangefinder post on the port side of the funnel.

The funnel cap and its platform with the anti-aircraft armament.

A section of the crane for hoisting ship boats and floatplanes.

Fully equipped battleship's midship section as seen during the Operation "Berlin".

The Flakvierling C/35 platform.

The floatplane on the catapult.

The catapult with the ship's floatplane in camouflage.

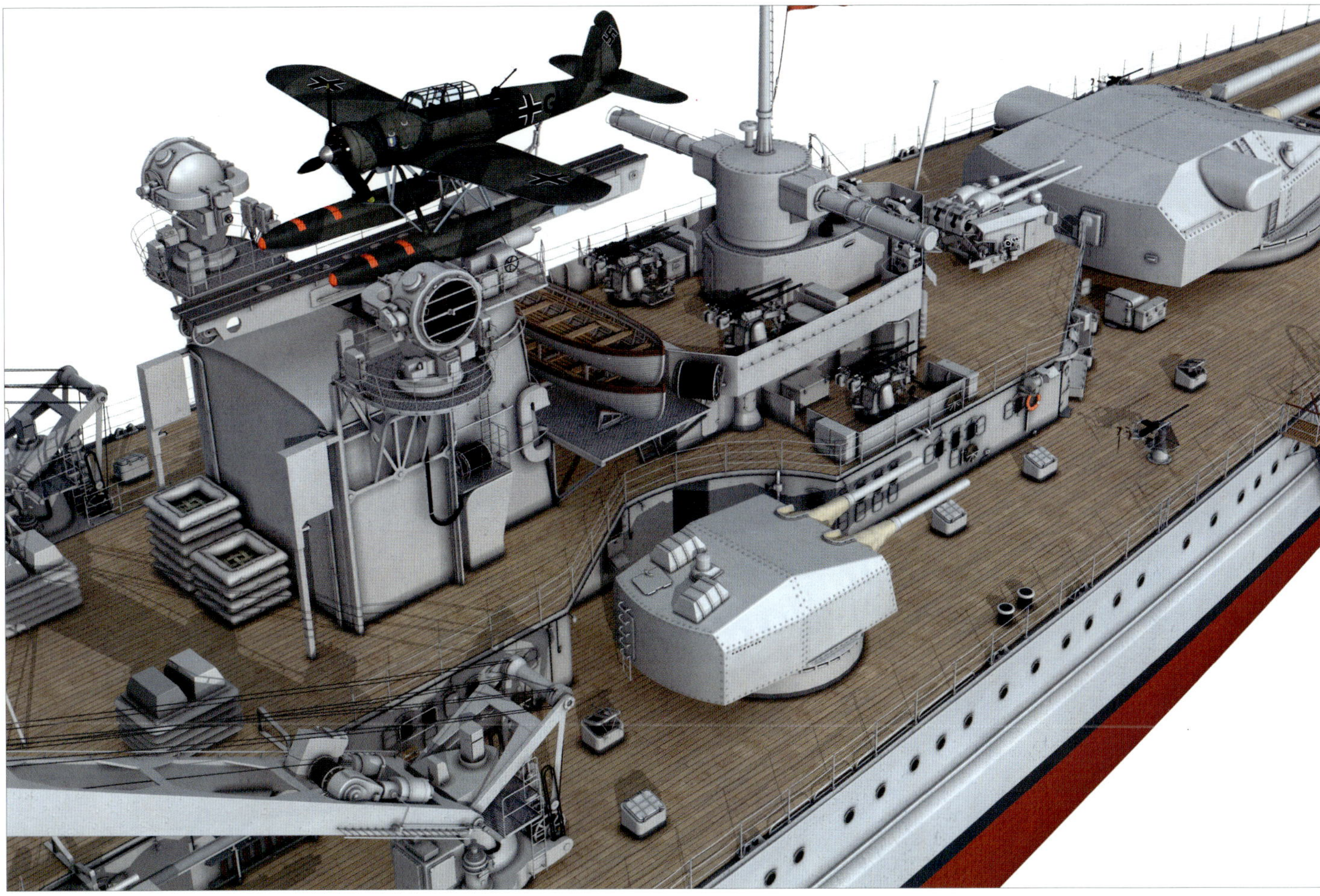

The stern section of the battleship with the floatplane on the catapult visible in the foreground.

The Arado floatplane of the catapult. Only a small bay that remained of the original hangar is visible in front of the catapult.

Details of the new small bay that was constructed after the battleship's refit.

A section of the port side of the lower superstructure with a gangway on the left.

The midship seen from astern. A 15 cm twin gun turret is visible in the foreground.

Section of the aft superstructure.

The battleship's forward superstructure with its fittings clearly visible.

A view of the small bay constructed in place of the original hangar.

The aft 10.5-metre base rangefinder.

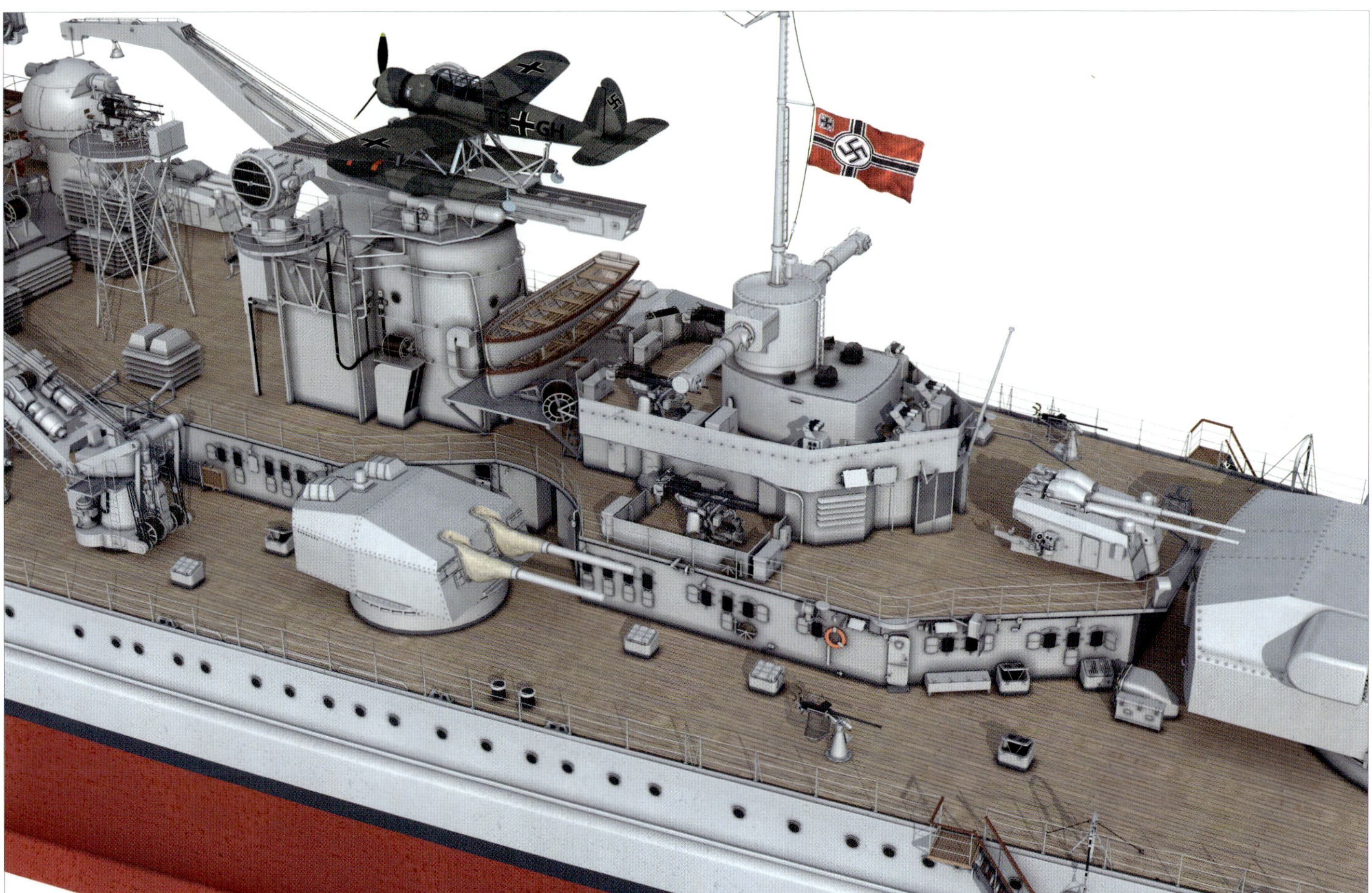

The anti-aircraft armament in the stern section of the battleship.

The ship's accommodation ladder.

The stern section of the battleship.

The main battery turret "C". In the initial period of her service there was a catapult installed on top of it, which considerably reduced its training speed. Therefore, it was removed in February 1940.

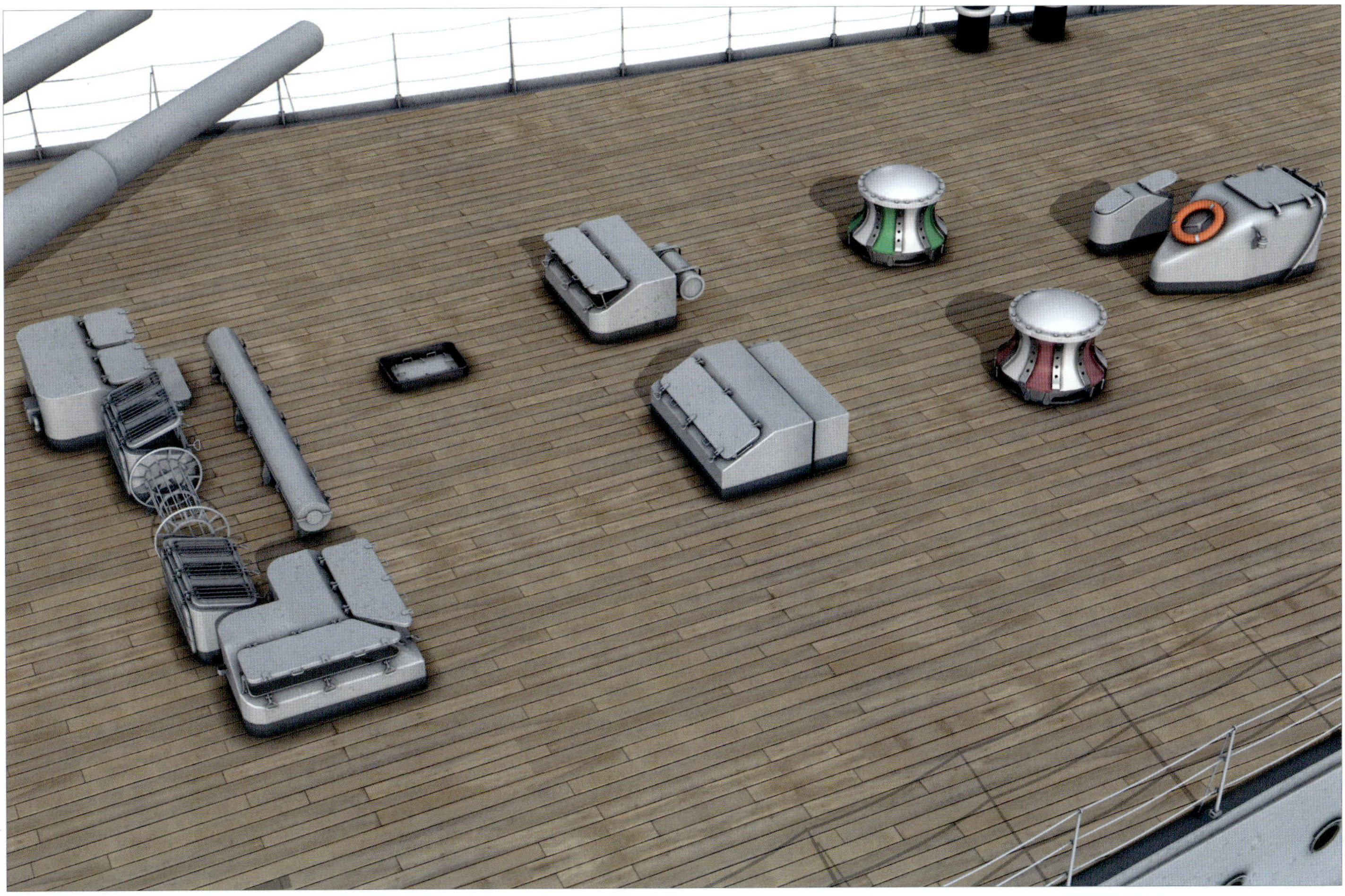

A section of the quarterdeck.

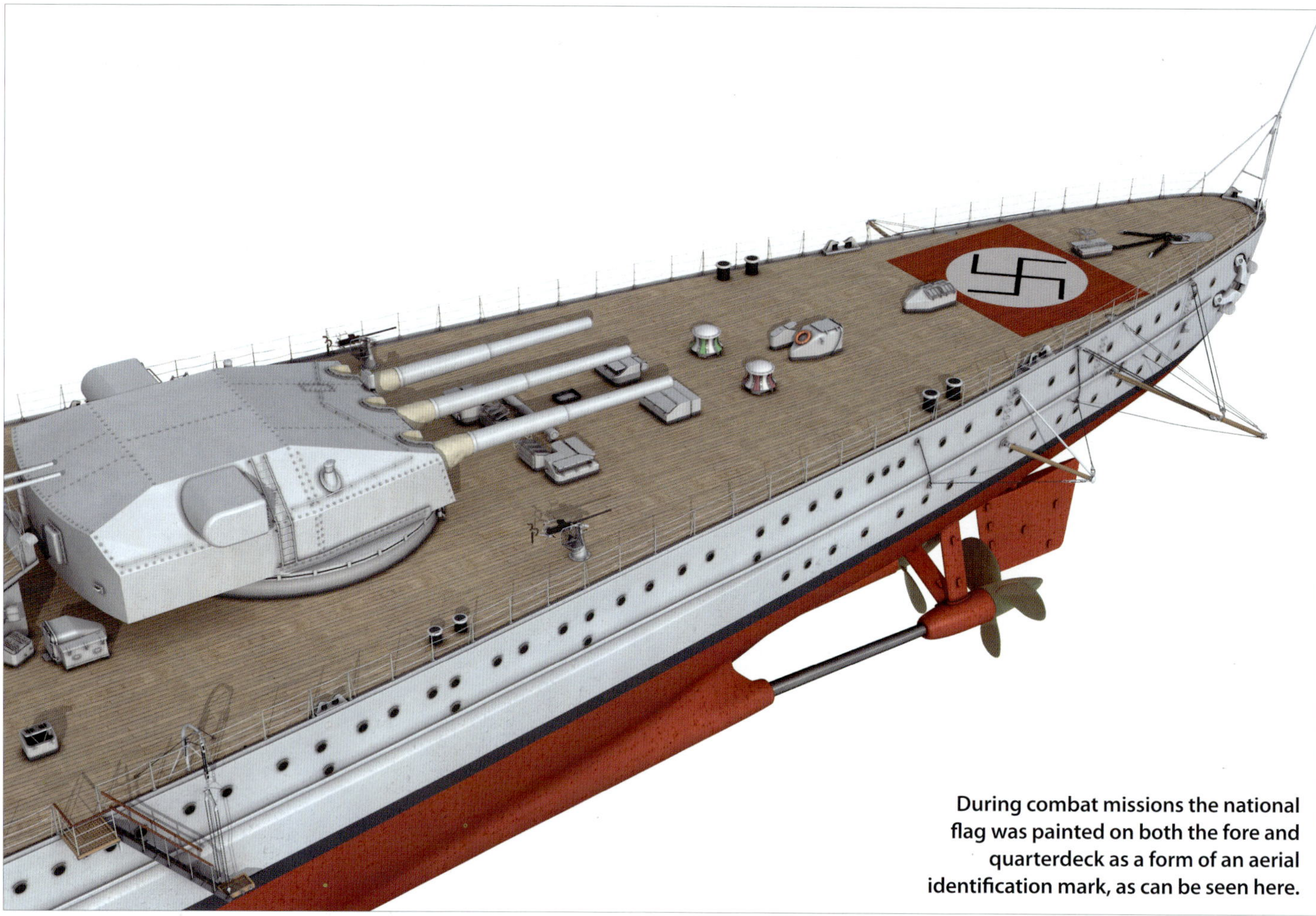

During combat missions the national flag was painted on both the fore and quarterdeck as a form of an aerial identification mark, as can be seen here.

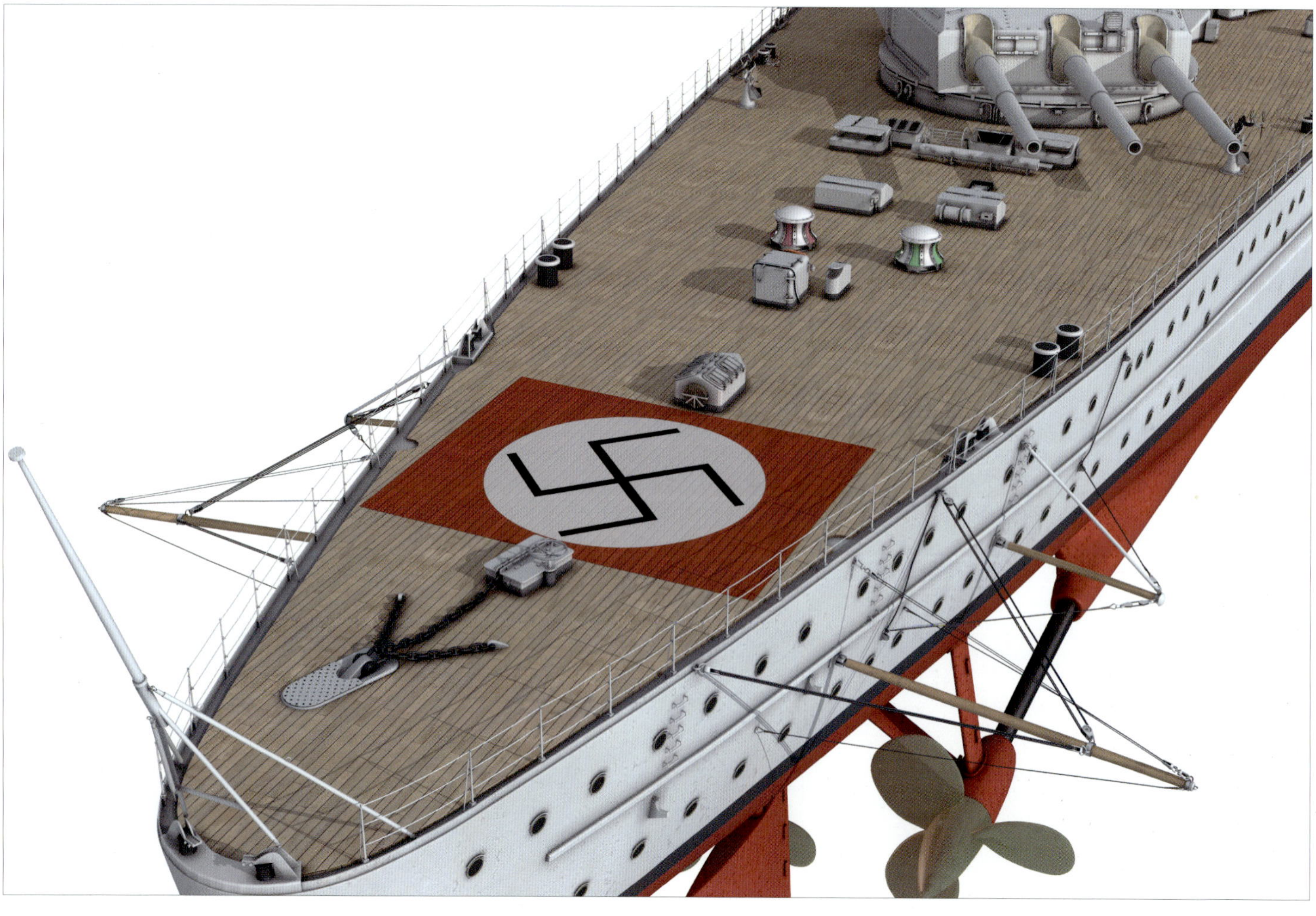

The battleship's port side seen from astern.

The main battery turret "C".

The aft superstructure with the anti-aircraft armament and optical fire control system visible.

The aft superstructure with its fittings.

The battleship's midship section with its fittings as seen during the Operation "Berlin".

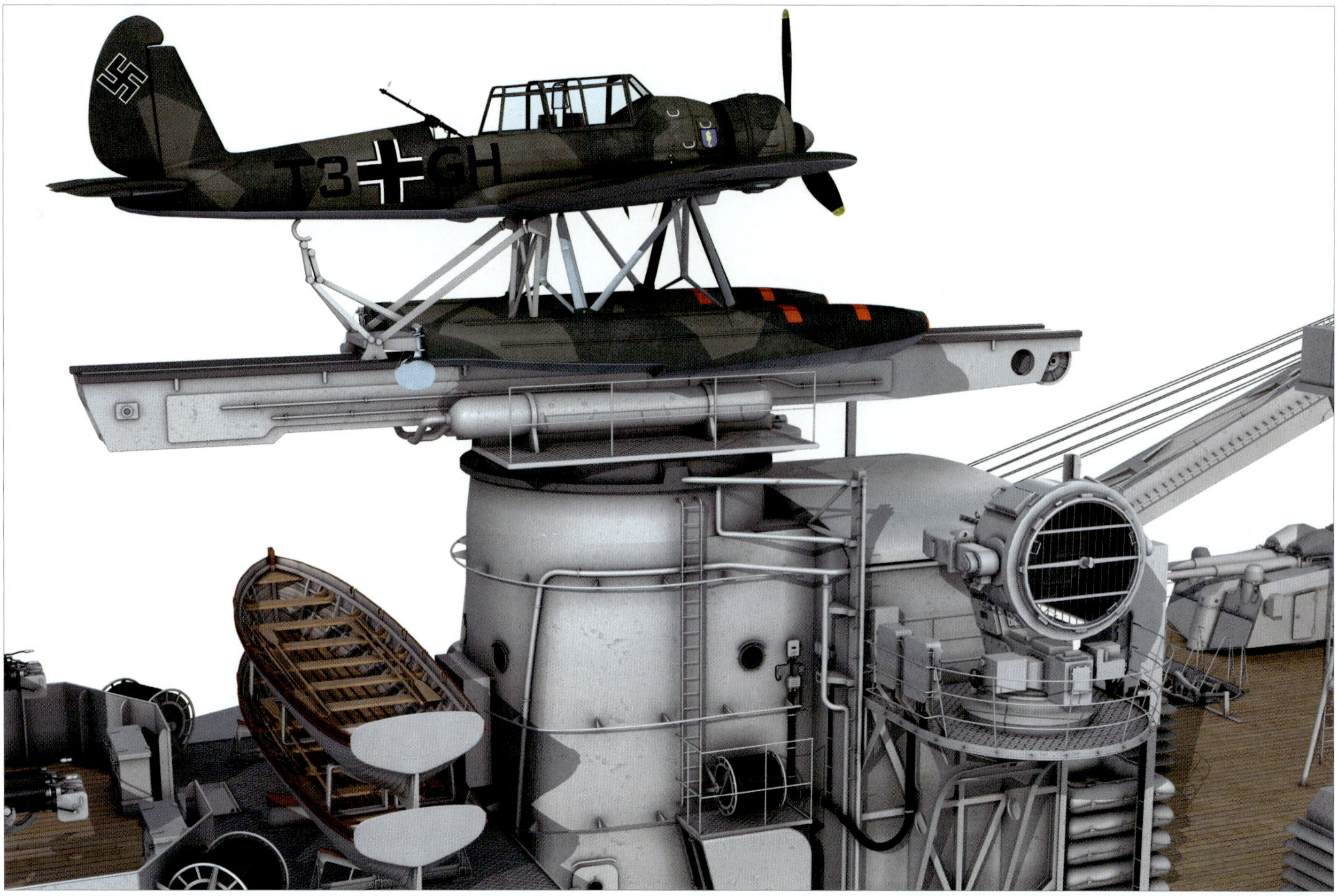

The battleship embarked Arado Ar 196 floatplanes, one of such planes is clearly visible on both illustrations.

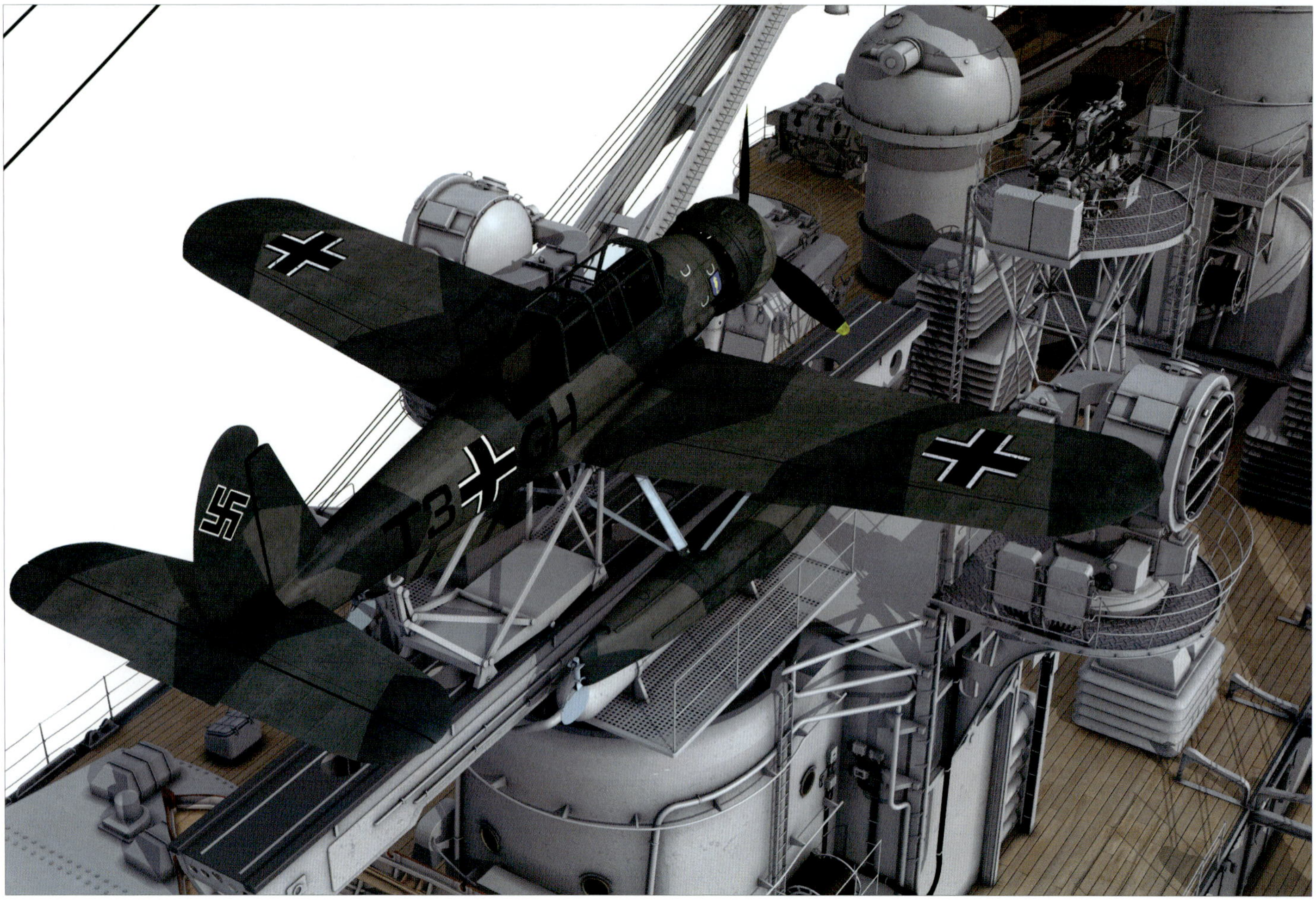

Details of the midship section around the small bay and the catapult.

Starboard side of the aft superstructure.

A view of the battleship's midship with a Flakvierling C/35 platform on the left. Both lattice cranes located on the sides of the hangar are partially visible.

A training 10.5 cm gun visible on the left was used for practice by the anti-aircraft battery crews.

A section of the lower superstructure in the vicinity of the funnel.

The battleship's midship section with its fittings.

The SL-6 rangefinder used for directing the 10.5 cm anti-aircraft battery fire.

Single 20 mm anti-aircraft cannons on the funnel's platform. Details of the mainmast attachment point are worthy of attention.

The battleship's midship section with its fittings.

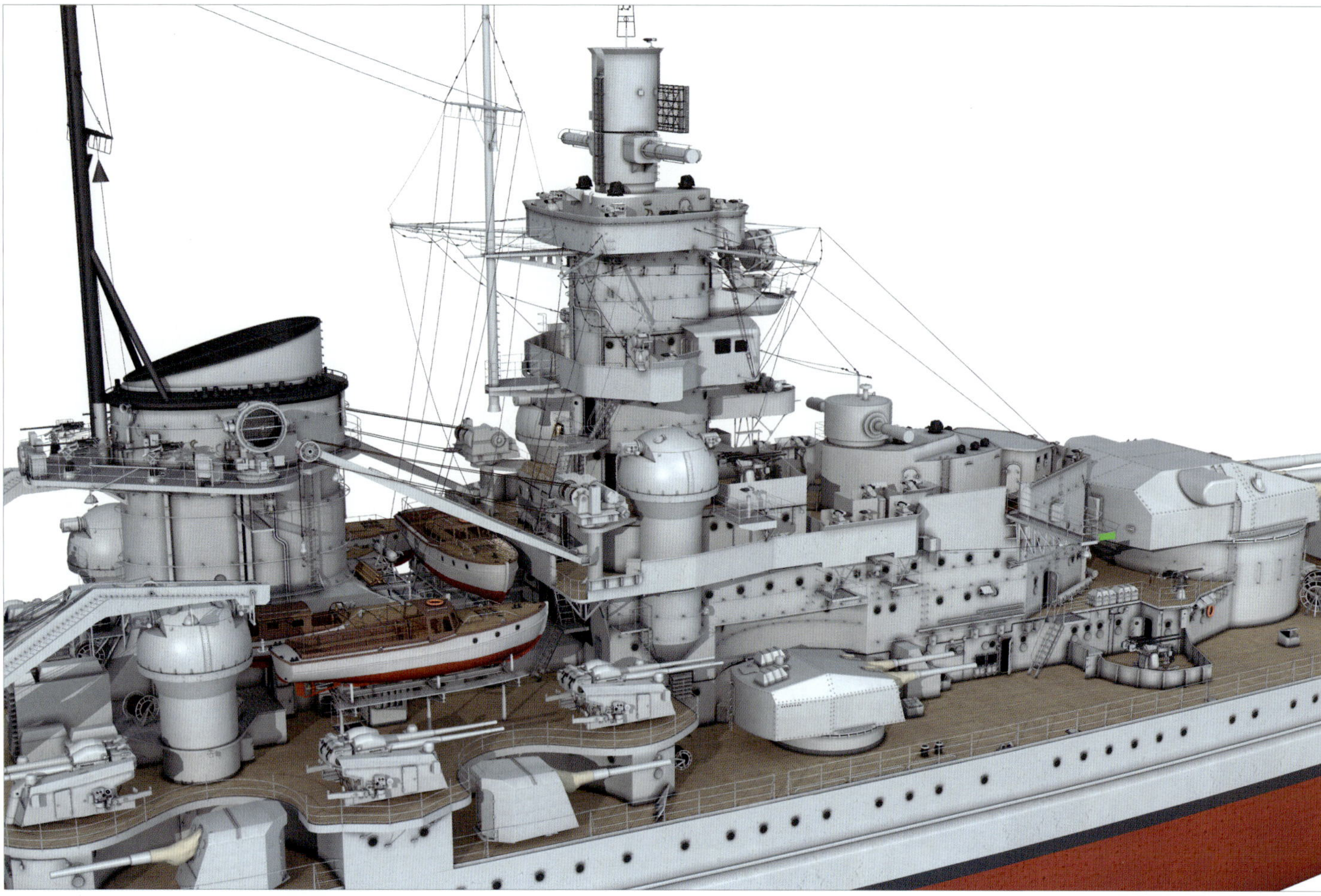

The bow section of the battleship with its fittings as seen during the Operation "Berlin".

Picket boats (Verkehrsboot) and Captain's gig (Admiralsboot) on chocks near the funnel.

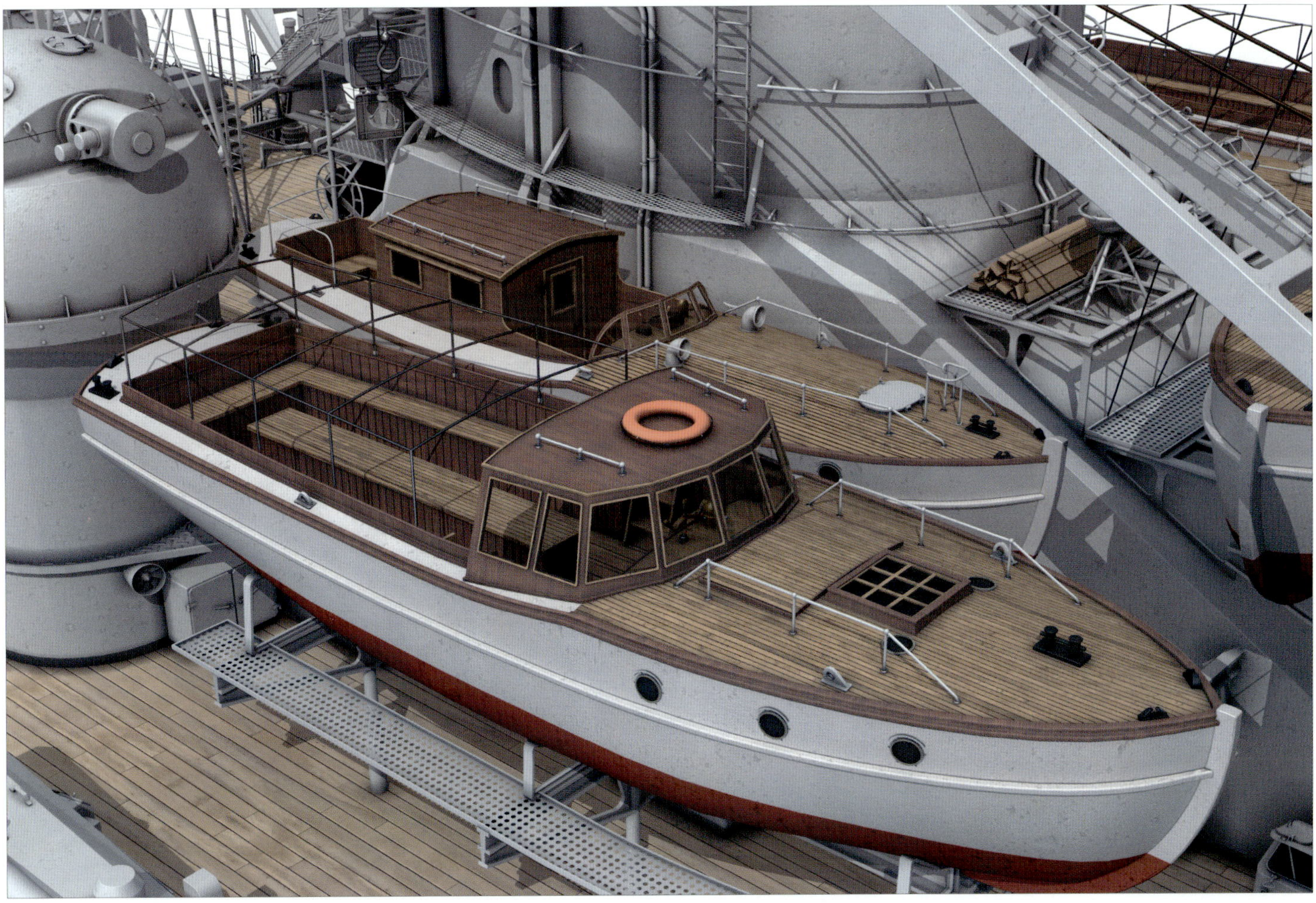

The forward superstructure with the cranes for hoisting picket boats visible in the foreground.

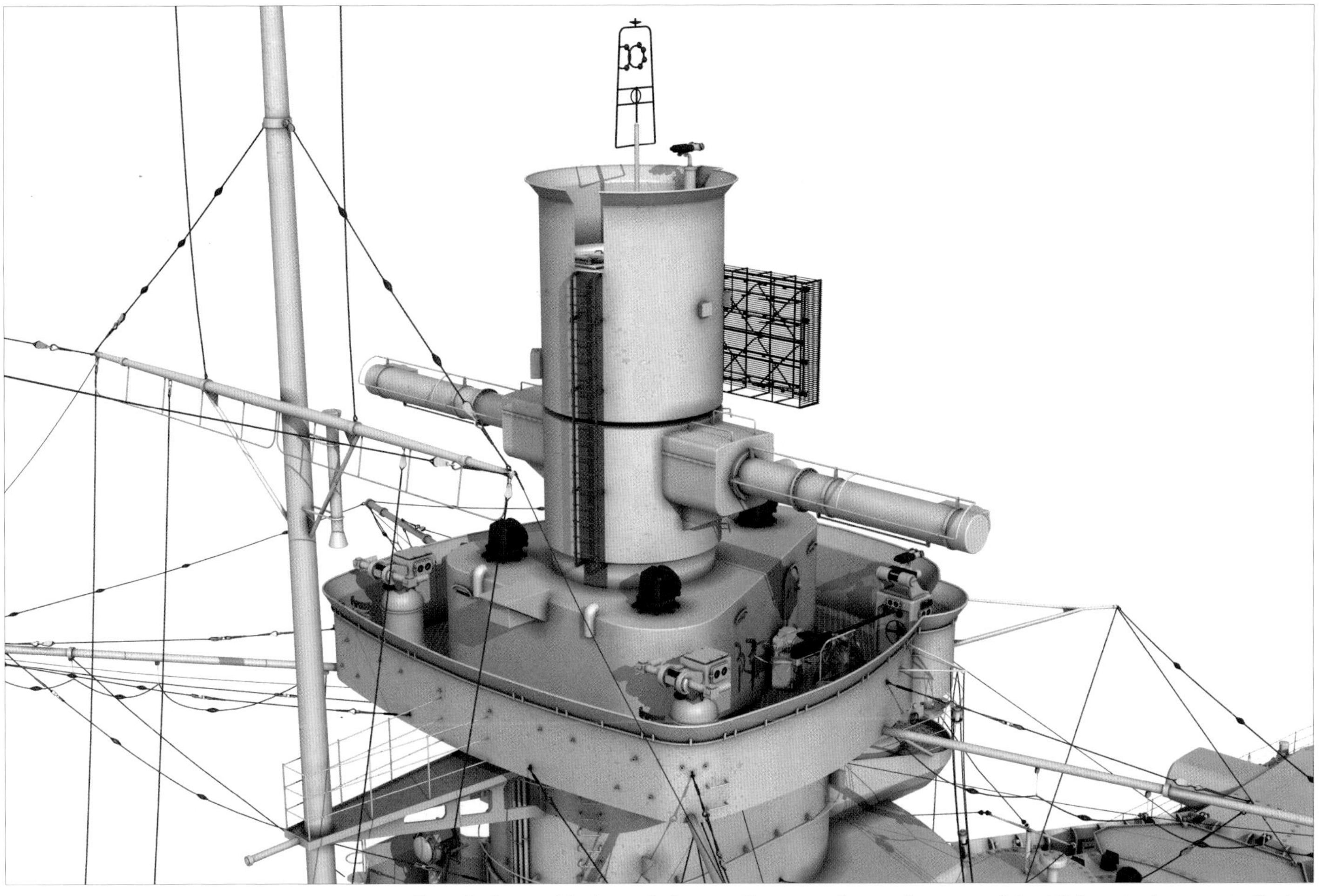

Top platform with the main battery fire control central room and the 10.5-metre base rangefinder.

The Admiral's bridge platform with its fittings seen from astern.

The secondary and anti-aircraft battery located around the forward superstructure.

A single 15 cm gun mounted in the midship section.

A section of the forward superstructure. The SL-6 rangefinder post is worty of attention. On the Gneisenau its base was conical, while the one on the Scharnhorst was square and flattened.

The floatplanes hoisting crane and the SL-6 rangefinder post.

The Admiral's bridge enclosed while the battleship was dry-docked in December 1939.

The forward superstructure.

A searchlight installed on a platform of the forward superstructure.

A view of the fittings located on the forward superstructure.

A starboard section of the forward superstructure located right behind the main battery turret "B".

The forward superstructure.

A section of the forward superstructure with a 37 mm cannon mount.

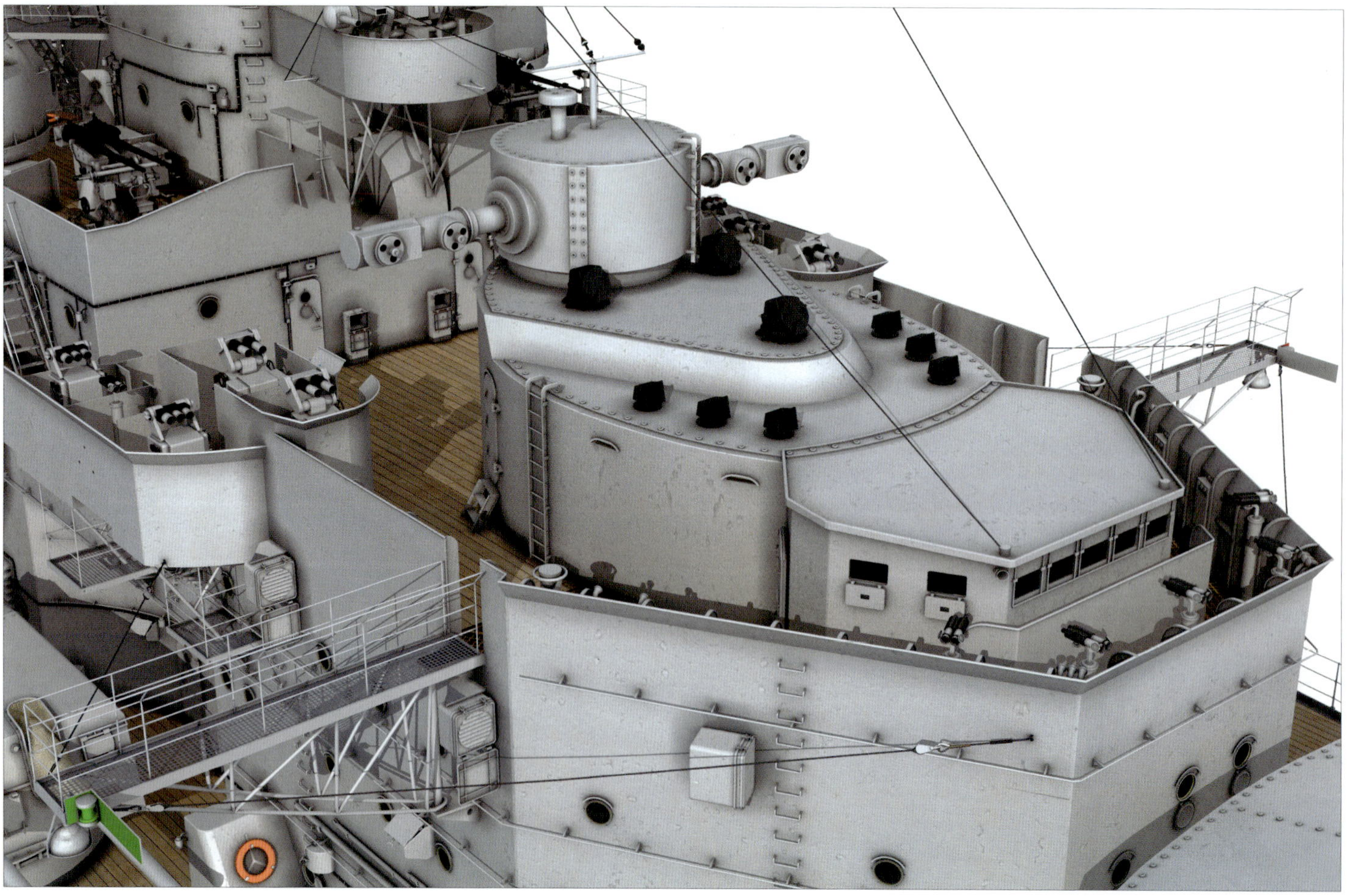

The control and manoeuvring bridge of the battleship with its fittings clearly visible.

The forward main battery turrets of the battleship. The turret "A" 10.5-metre rangefinder was removed during her stay at Brest and its openings were closed.

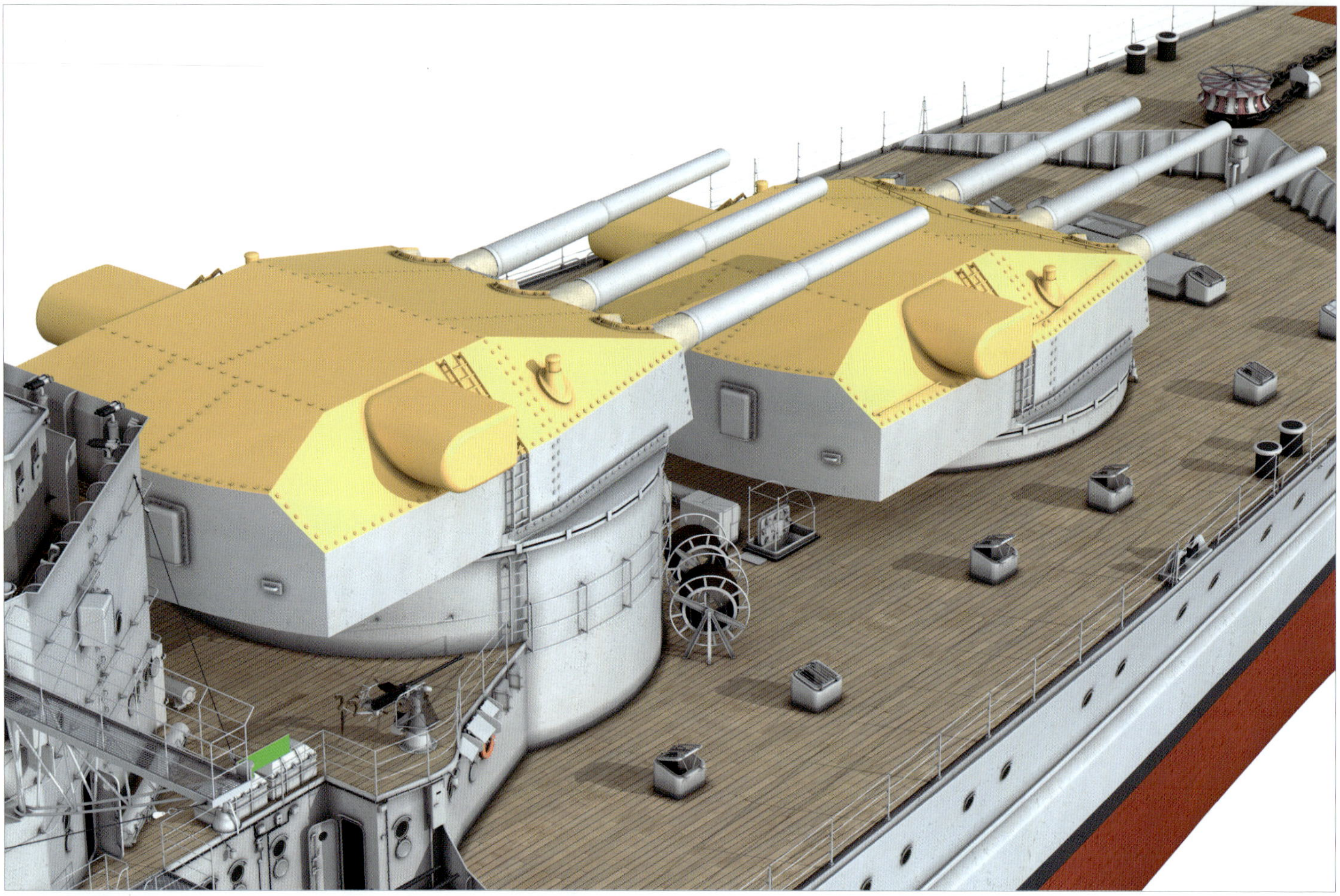

During the Operation "Berlin" the battleship's turrets were painted yellow for aerial identification purposes.

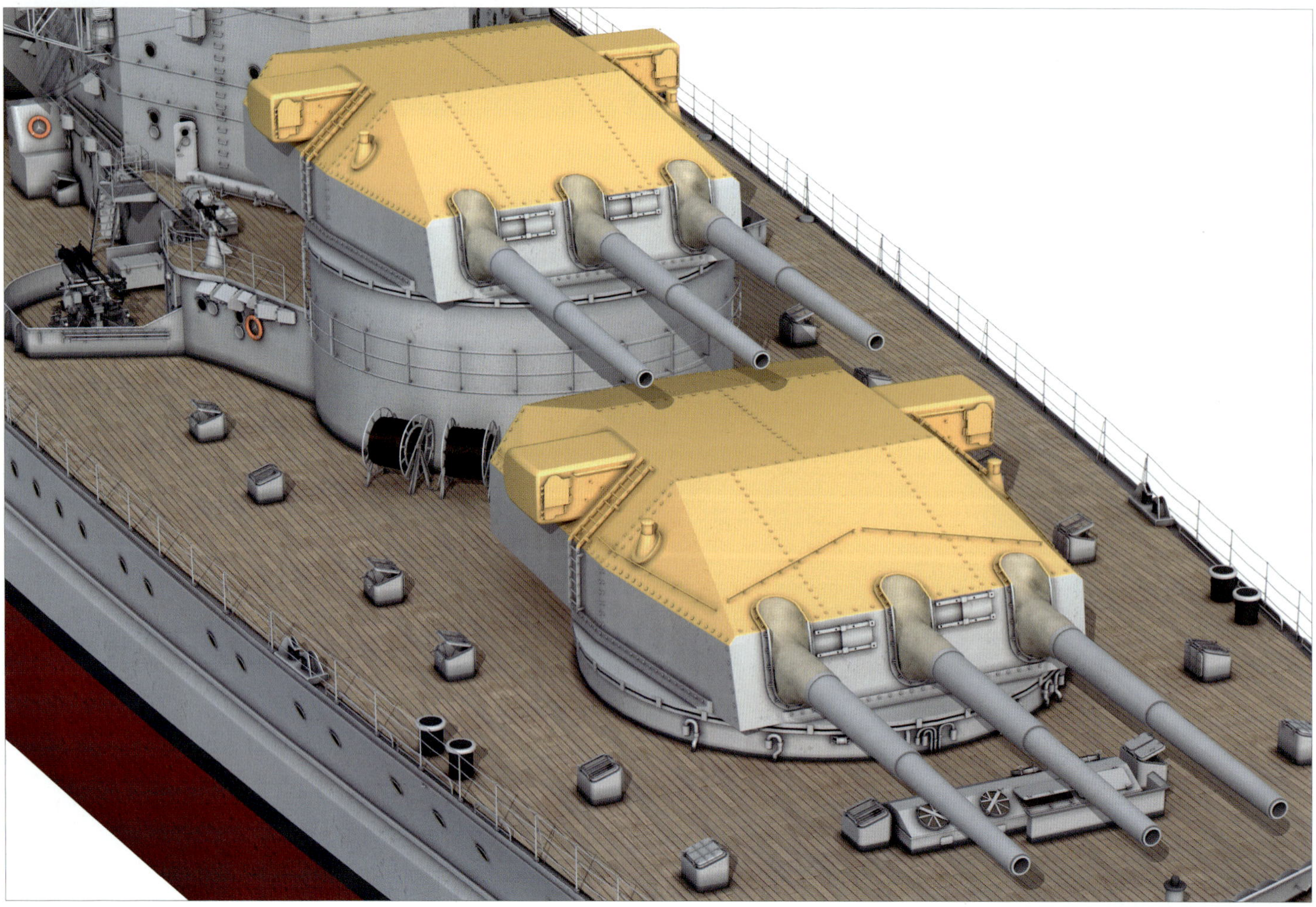

The main battery turrets were painted standard light gray during some periods of the battleship's service. This painting scheme was supplemented with other identification markings like white circles, red filled circles and white stripes depending on individual combat missions.

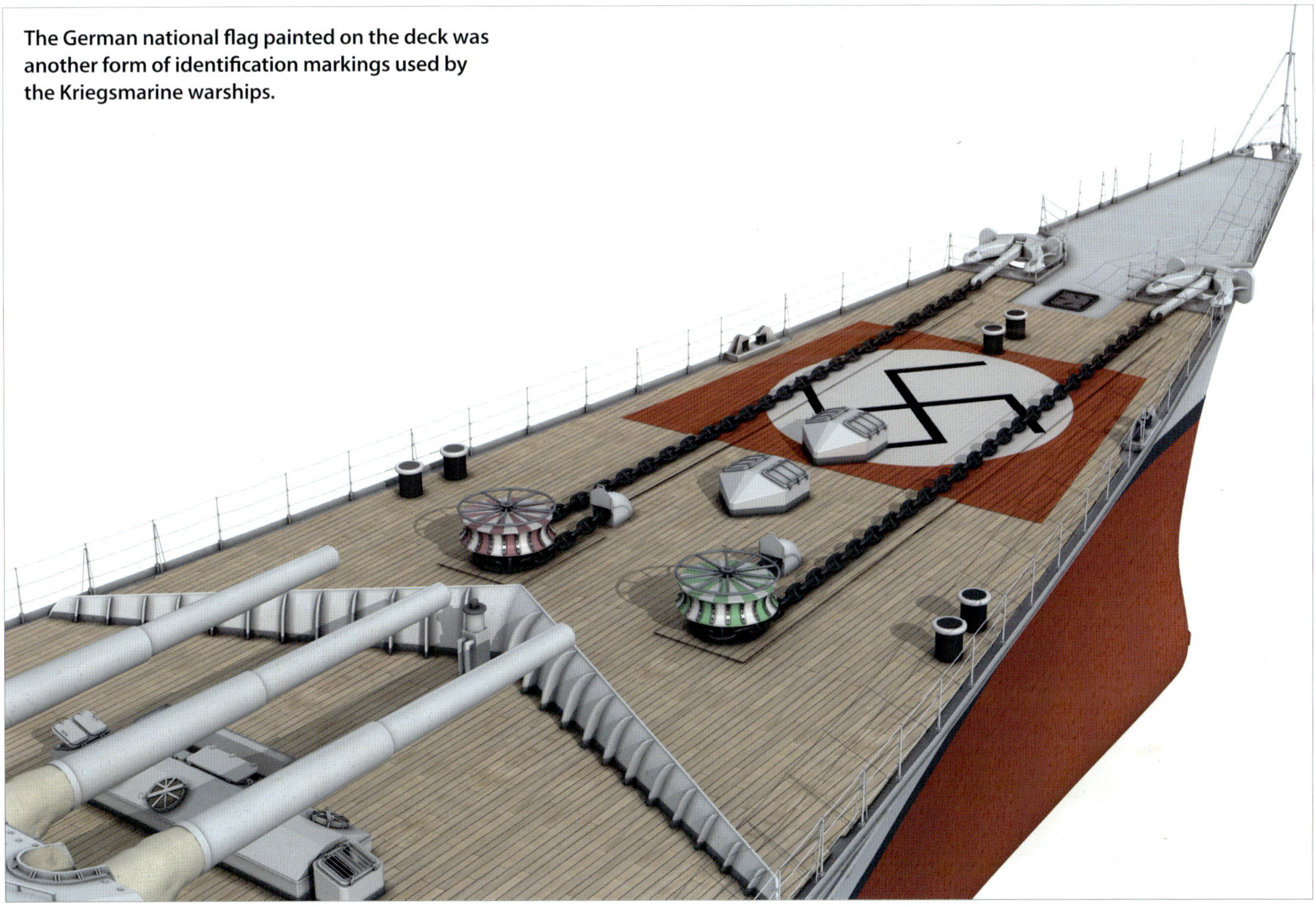

The German national flag painted on the deck was another form of identification markings used by the Kriegsmarine warships.

The main battery turret "B" and the forward 37 mm cannon mount.

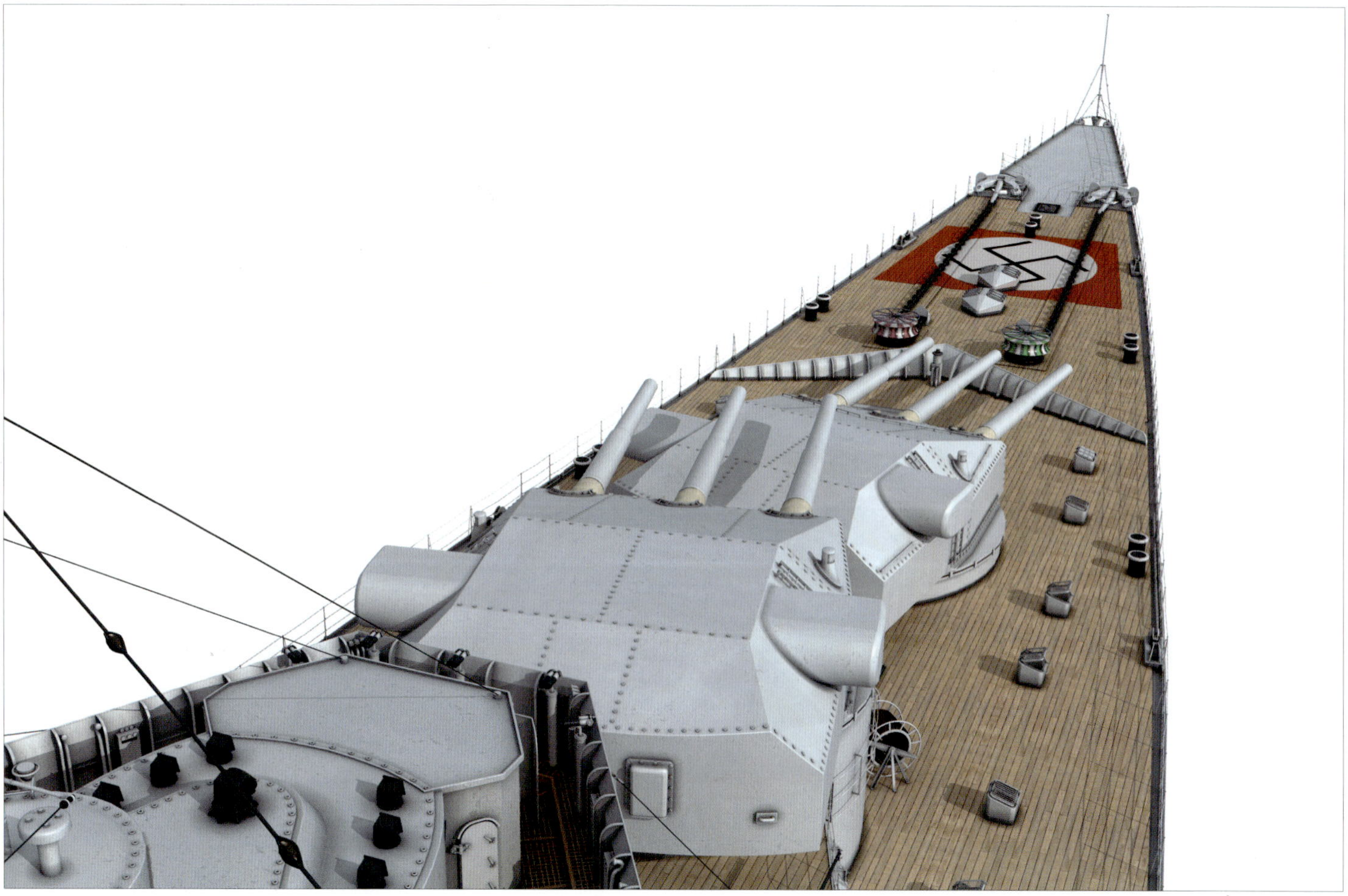

The main battery turrets "A" and "B".

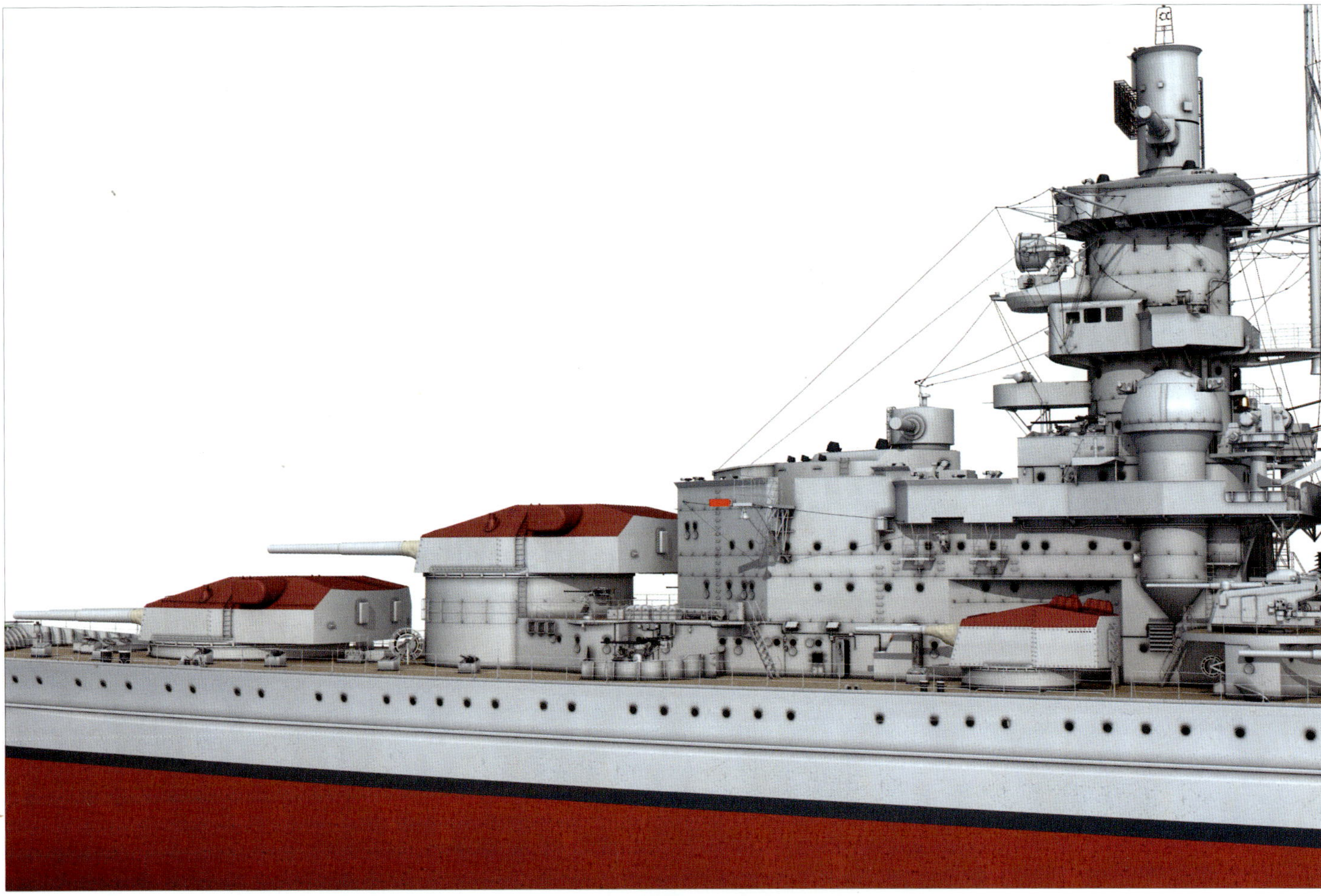

The identification markings on the main battery turrets "A" and "B" in the final days of the Operation "Wesserübung".

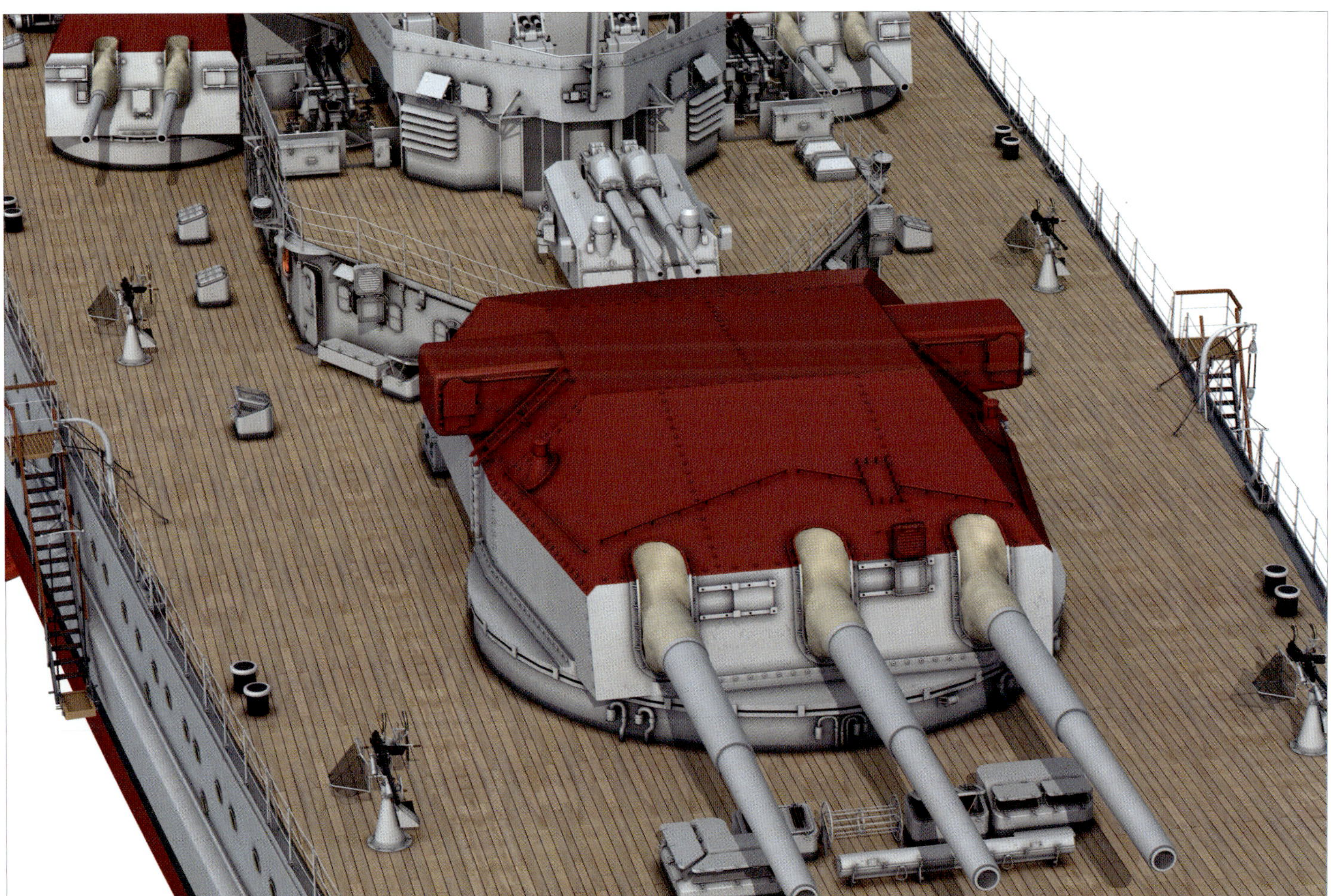

The secondary battery of the battleship consisted of 150 mm guns mounted in four twin LC/34 turrets seen in these illustrations.

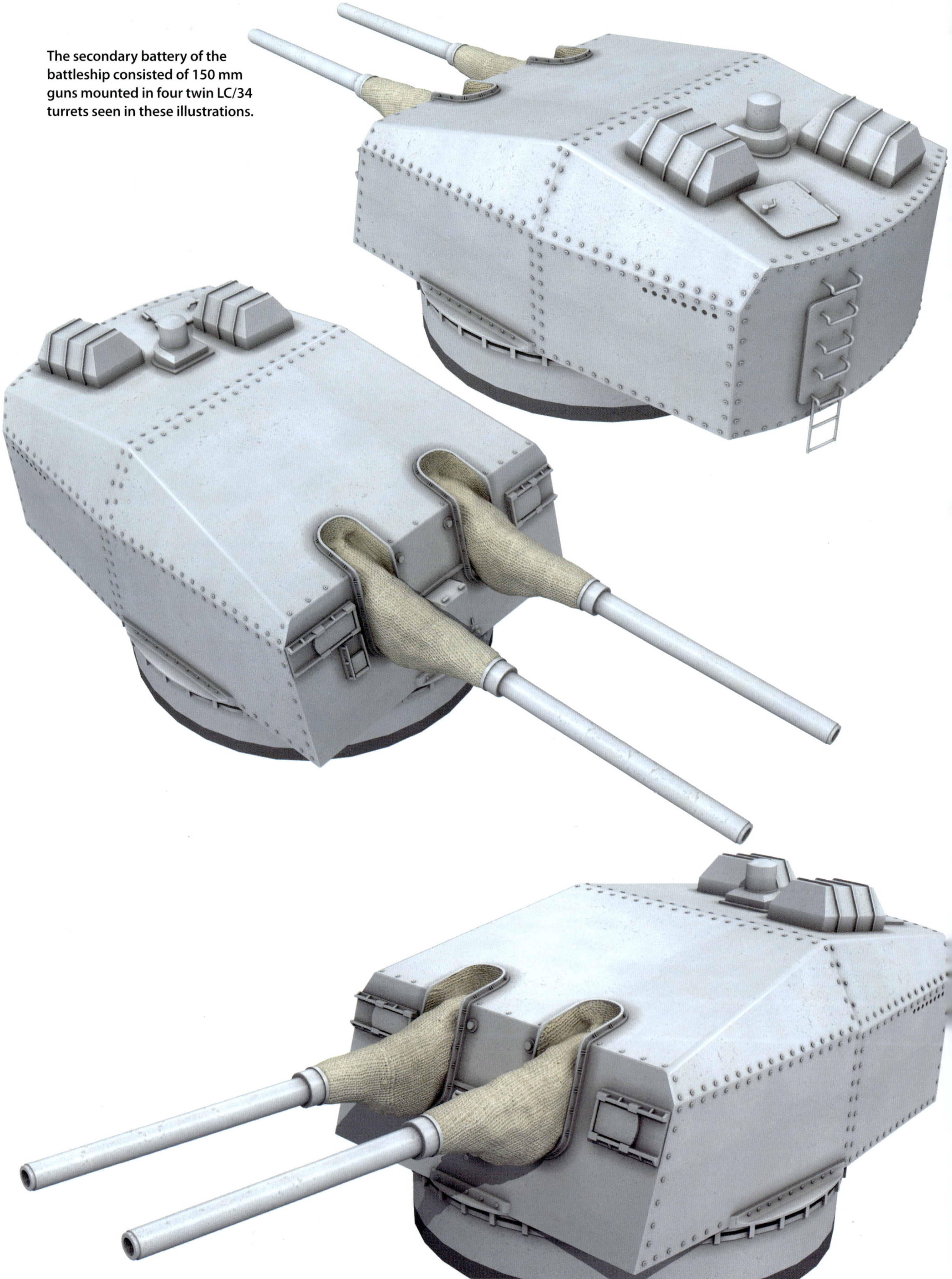

The main battery turrets "A" and "B" of the battleship *Gneisenau*. The top and middle illustration shows them in the standard light grey painting scheme, while the bottom one shows them with dark grey tops.

The main battery turret "C" already without the catapult.

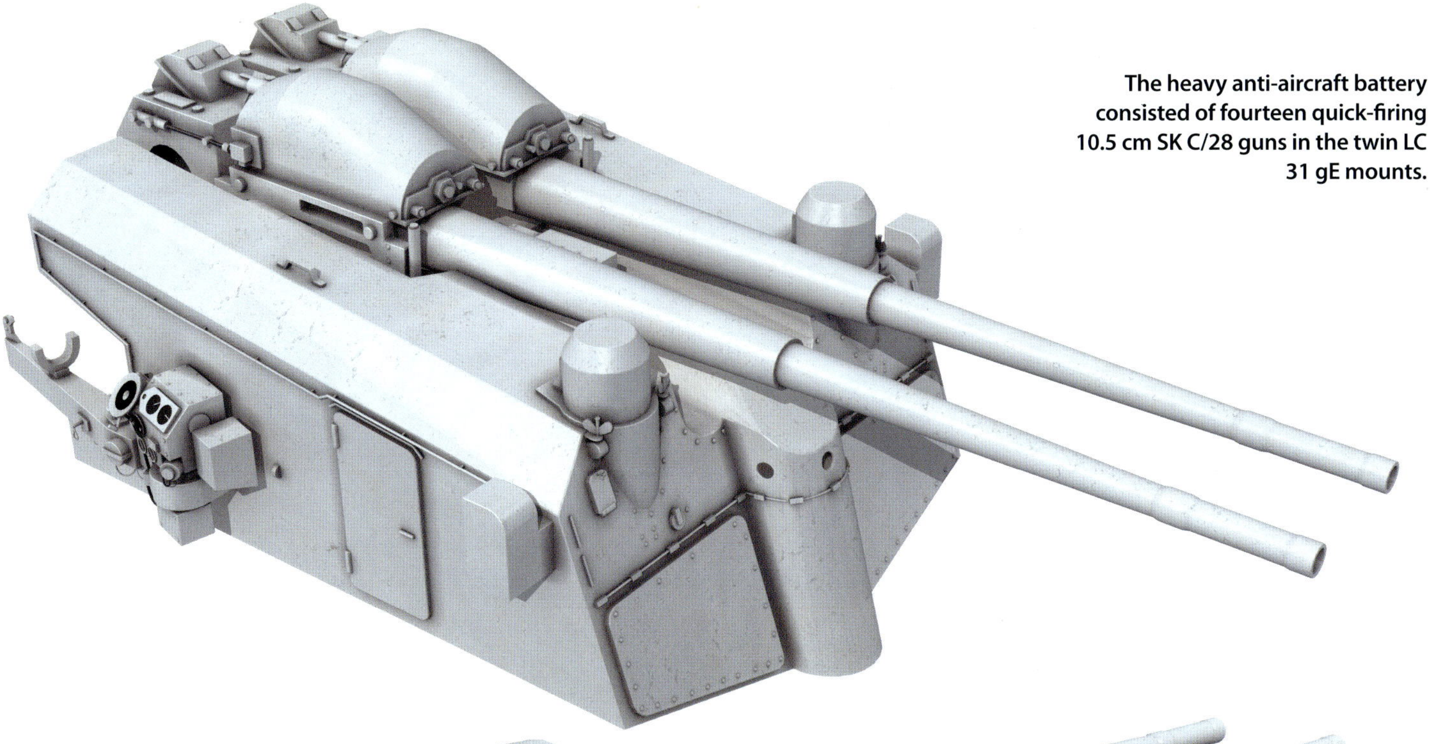

The heavy anti-aircraft battery consisted of fourteen quick-firing 10.5 cm SK C/28 guns in the twin LC 31 gE mounts.

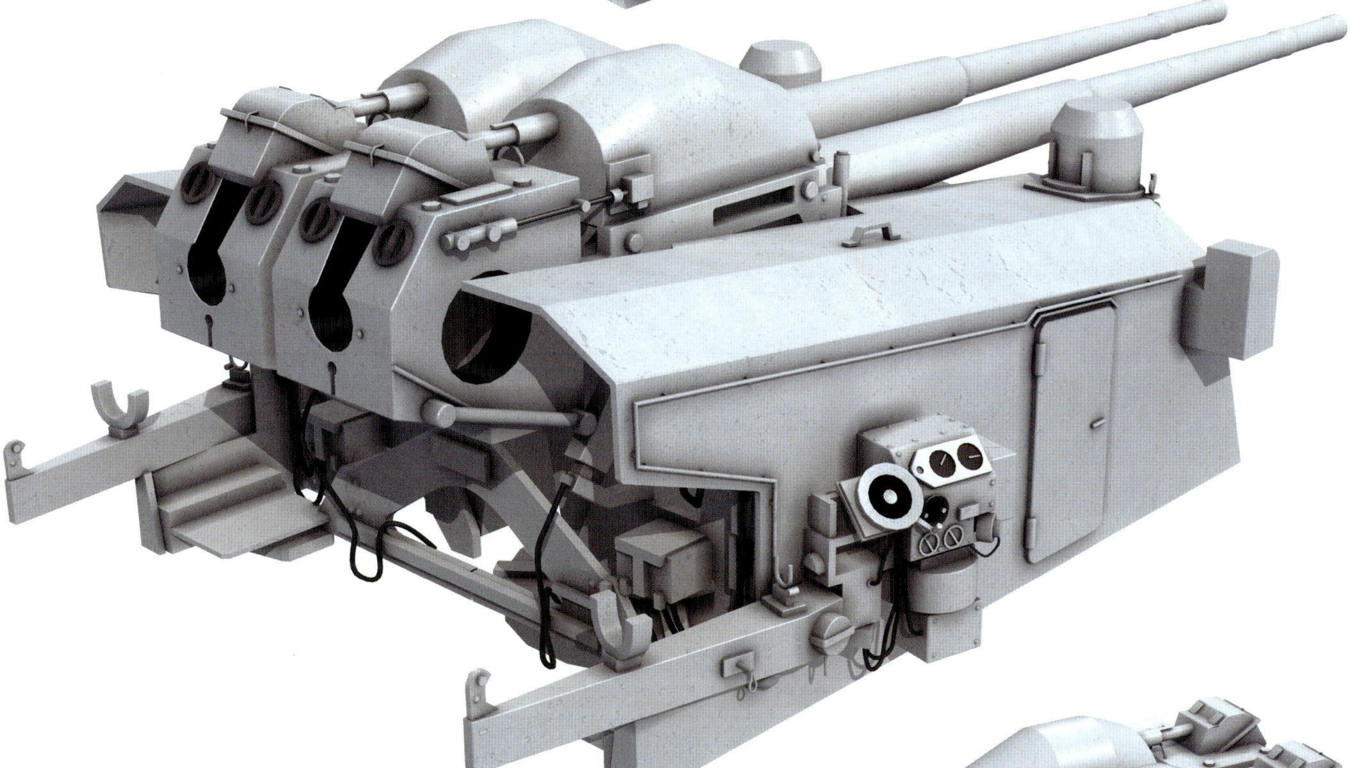

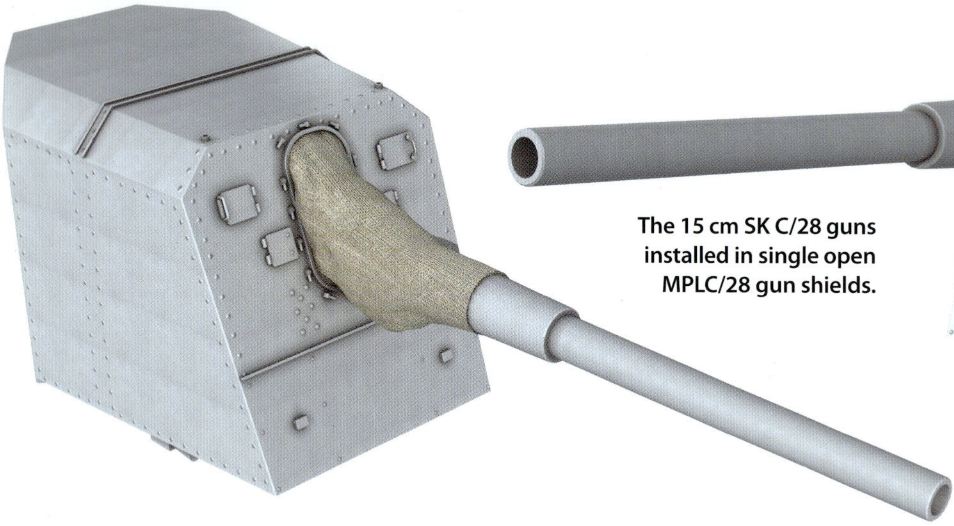
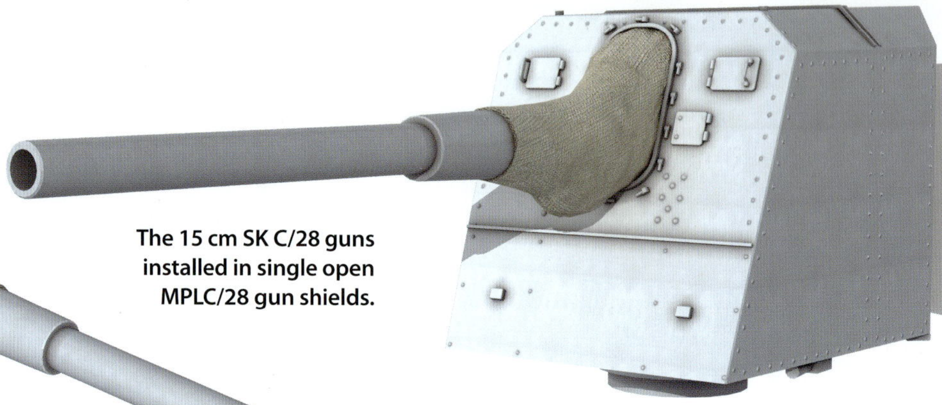

The 15 cm SK C/28 guns installed in single open MPLC/28 gun shields.

Single 20 mm MG c/30 cannons constituted the battleship's light anti-aircraft battery.

Medium anti-aircraft battery consisted of sixteen 37 mm cannons in eight twin C/30 mounts.

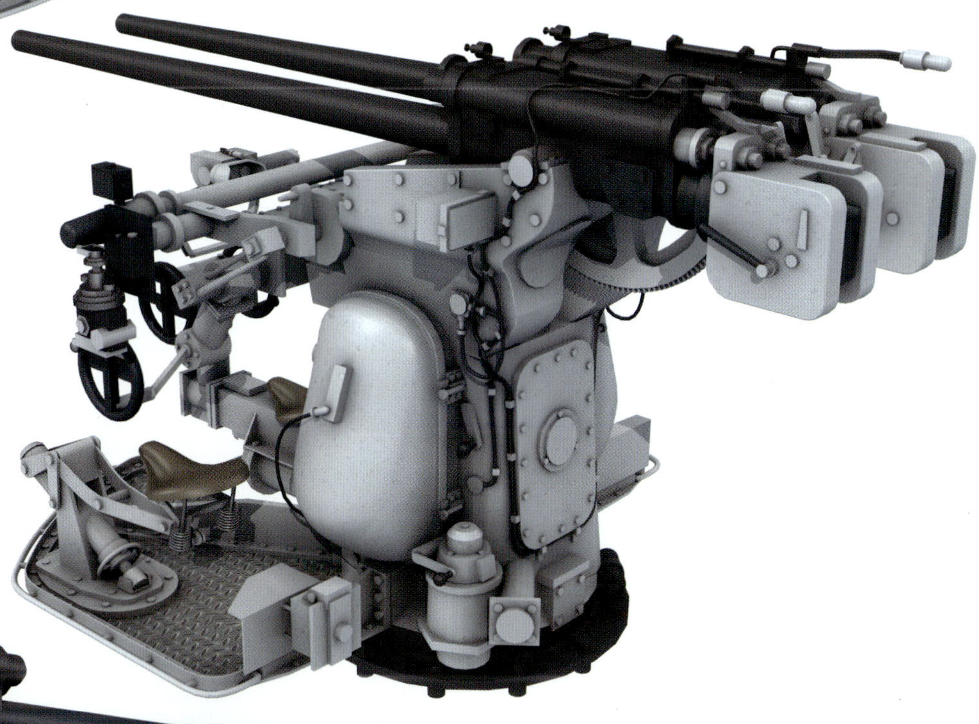

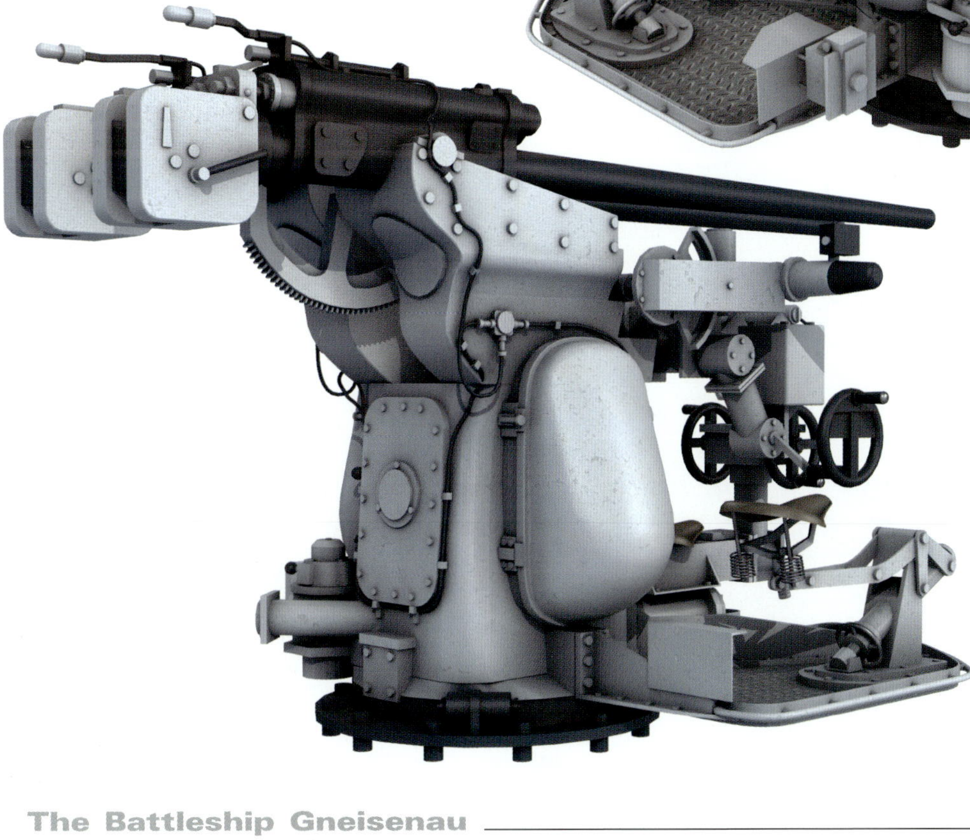

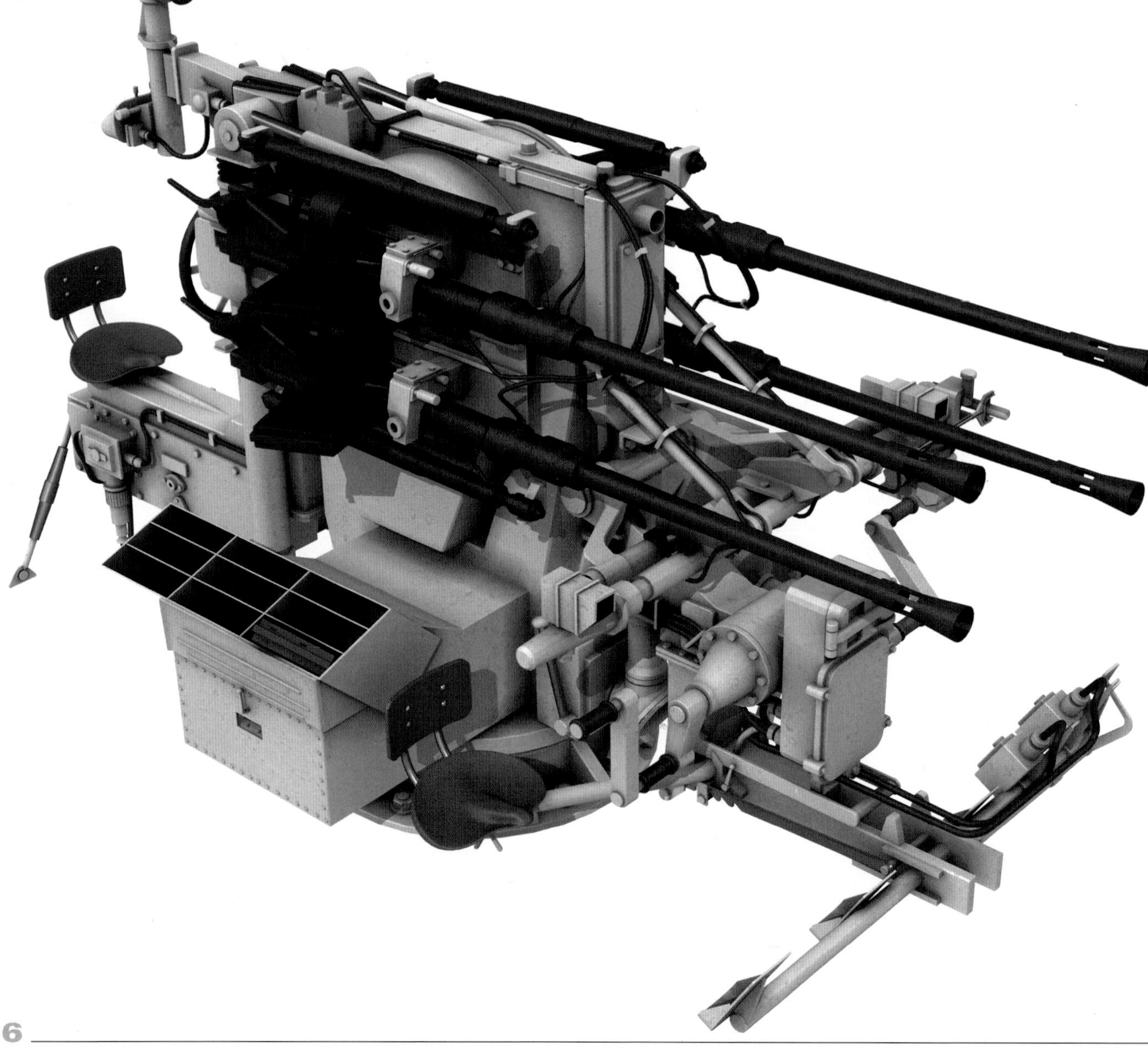

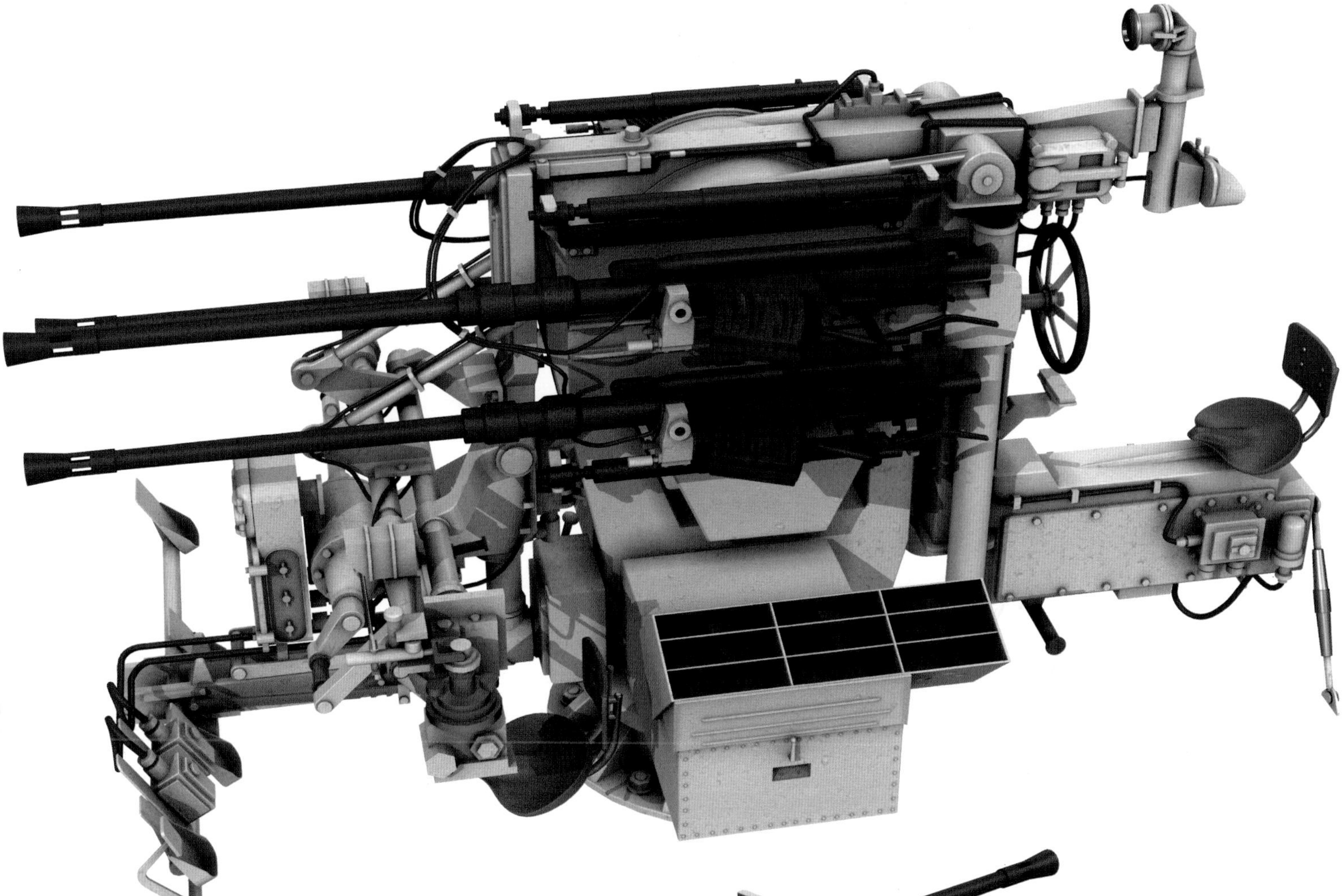

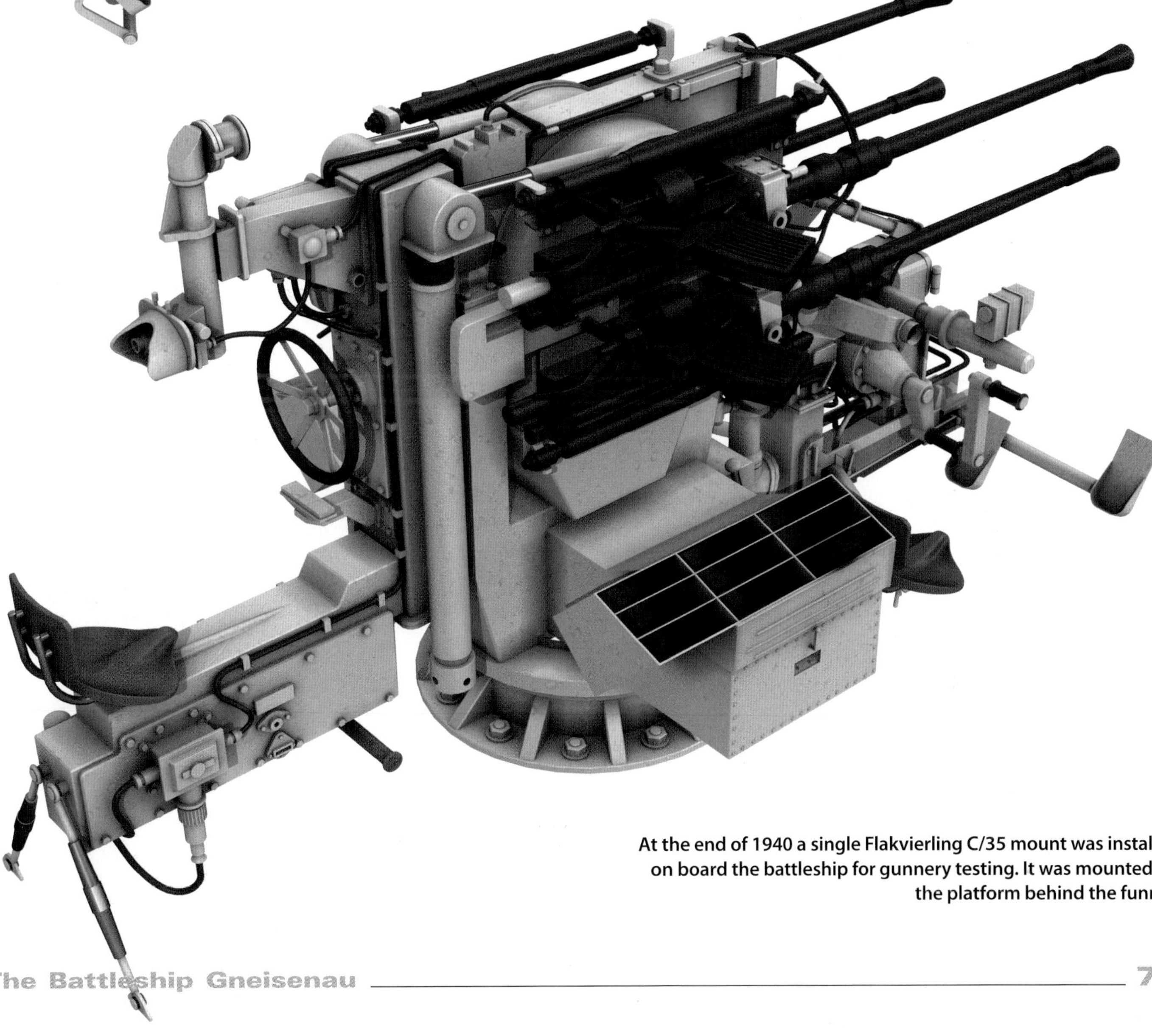

At the end of 1940 a single Flakvierling C/35 mount was installed on board the battleship for gunnery testing. It was mounted on the platform behind the funnel.

Bow view of the battleship *Gneisenau* at the beginning of 1941.

Stern view of the battleship *Gneisenau* at the beginning of 1941.

Starboard side view of the battleship *Gneisenau*.

Battleship Gneisenau – Mariusz Motyka • Mirosław Skwiot • First edition • LUBLIN 2012 • ISBN 978-83-62878-28-4

• Translation: **Kazimierz Zygadło** • Cover artwork: **Mariusz Motyka** • Description of images: **Mirosław Skwiot** • Design: KAGERO STUDIO

Oficyna Wydawnicza KAGERO • www.kagero.pl • e-mail: kagero@kagero.pl, marketing@kagero.pl

Editorial office, Marketing, Distribution: KAGERO PUBLISHING Sp. z o.o., ul. Mełgiewska 9F, 20-209 Lublin, phone (+48) 81 749 20 20, fax (+48) 81 749 11 81

w w w . k a g e r o . p l